Learning from the Ground Up

Learning from the Ground Up

Global Perspectives on Social Movements and Knowledge Production

Edited by
Aziz Choudry and Dip Kapoor

LEARNING FROM THE GROUND UP
Copyright © Aziz Choudry and Dip Kapoor, 2010.
All rights reserved.

First published in 2010 by
PALGRAVE MACMILLAN®
in the United States—a division of St. Martin's Press LLC,
175 Fifth Avenue, New York, NY 10010.

Where this book is distributed in the UK, Europe and the rest of the world, this is by Palgrave Macmillan, a division of Macmillan Publishers Limited, registered in England, company number 785998, of Houndmills, Basingstoke, Hampshire RG21 6XS.

Palgrave Macmillan is the global academic imprint of the above companies and has companies and representatives throughout the world.

Palgrave® and Macmillan® are registered trademarks in the United States, the United Kingdom, Europe and other countries.

ISBN: 978–0–230–62103–9

Library of Congress Cataloging-in-Publication Data

    Learning from the ground up : global perspectives on social movements and knowledge production / edited by Aziz Choudry and Dip Kapoor.
      p. cm.
    ISBN 978–0–230–62103–9 (alk. paper)
    1. Social movements—Study and teaching. 2. Social action—Study and teaching. 3. Action research. 4. Experiential learning. I. Choudry, A. A. II. Kapoor, Dip.

HM881.L427 2010
303.48'401—dc22
                                      2010009106

A catalogue record of the book is available from the British Library.

Design by Newgen Imaging Systems (P) Ltd., Chennai, India.

First edition: September 2010

10 9 8 7 6 5 4 3 2 1

Printed in the United States of America.

I dedicate this book to all who mobilize, organize, educate, and agitate for self-determination, justice, and dignity, and I gratefully acknowledge the Indigenous Peoples of Aotearoa and Turtle Island for providing me with the spaces to think and write.

—*Aziz*

I dedicate this work to the Dalit victims of Kandhamal, and to Dalit struggles for dignity, equality, and political-economic justice in India.

—*Dip*

Contents

Acknowledgments ix

1 Learning from the Ground Up: Global Perspectives on Social Movements and Knowledge Production 1
  Aziz Choudry and Dip Kapoor

Part I Making Knowledge and Learning from the Politics of Knowledge Production and Representation: "Civil Society," Academe, and Social Activism—Tensions, Challenges, and Dilemmas

2 Global Justice? Contesting NGOization: Knowledge Politics and Containment in Antiglobalization Networks 17
  Aziz Choudry

3 Knowledge and Power in South Africa: Xenophobia and Survival in the Post-Apartheid State 35
  Ashwin Desai and Shannon Walsh

4 On the Question of Expertise: A Critical Reflection on "Civil Society" Processes 53
  Robyn Magalit Rodriguez

5 Whatever Happened to the Counter-Globalization Movement? Some Reflections on Antagonism, Vanguardism, and Professionalization 69
  Kees Hudig and Emma Dowling

6 Collective Approaches to Activist Knowledge: Experiences of the New Anti-Apartheid Movement in Toronto 85
  Rafeef Ziadah and Adam Hanieh

7 The Subjectivation of Marriage Migrants in Taiwan: The
   Insider's Perspectives   101
   Hsiao-Chuan Hsia

Part II Making Knowledge and Learning from Unions, Worker Alliances, and Left Party-Political Activism

8 Learning to Win: Exploring Knowledge and Strategy
   Development in Anti-Privatization Struggles in Colombia   121
   Mario Novelli

9 Worker Education and Social Movement Knowledge
   Production: Practical Tensions and Lessons   139
   David Bleakney and Michael Morrill

10 Conversations on the M60: Knowledge Production through
   Collective Ethnographies   157
   Biju Mathew

11 Vanguards and Masses: Global Lessons from the Grenada
   Revolution   173
   David Austin

Part III Making Knowledge and Learning from Peasant and Indigenous Peoples' Struggles

12 Learning and Knowledge Production in Dalit Social
   Movements in Rural India   193
   Kumar Prasant and Dip Kapoor

13 *Anjuman-e-Mazareen Punjab*: Ownership or Death—The
   Struggle Continues   211
   Azra Talat Sayeed and Wali Haider

14 How Do You Say *Netuklimk* in English? Using
   Documentary Video to Capture Bear River First Nation's
   Learning through Action   227
   Martha Stiegman and Sherry Pictou

Notes on Contributors   243

Index   249

Acknowledgments

Dominant academic norms tend to encourage the reproduction of an individualistic and competitive pursuit of research and knowledge organized in the interests of what Dorothy Smith refers to as *ruling relations*. Yet this book embodies our firm conviction in and recognition of the conceptual resources and knowledge produced in people's collective struggles against injustice, not to mention the social and academic significance of collaborative efforts such as this; an effort that would not have transpired without the commitment and enthusiasm of all the contributors. Most contributors are located within, or work closely with, a range of anti-colonial and anti-imperialist social movements, communities, and popular mobilizations in varied contexts in the Asia-Pacific, Africa, Europe, the Americas, and the Caribbean, as well as within regional and global movement networks. Our sincere and heartfelt thanks goes to all of them given the challenging task of balancing academic, activist, and other aspects of their lives while remaining engaged on the frontlines of struggles in diverse and difficult contexts, wherein writing book chapters may not be a priority. Yet all remained committed to this project from the beginning and through each successive phase. Given the level of this commitment, we sincerely hope that this book succeeds in making a contribution toward these political struggles, while adding to the scholarly literature on knowledge production and social movements.

Our deep gratitude goes to Alison Crump for her editorial support and her professionalism, efficiency, and patience in helping us prepare this book. We are indebted to her for her unflagging attention to detail and hard work in relation to formatting and referencing, and in keeping the communication going between all parties concerned. Thanks, Alison!

For the striking cover art, we are indebted to U.S.-based activist and photojournalist Orin Langelle, codirector of Global Justice Ecology Project; more of Orin's excellent work can be viewed at http://globaljusticeecology.org.

Many thanks are due to Julia Cohen and Samantha Hasey at Palgrave Macmillan for their enthusiastic support and assistance in seeing this book project through to fruition, and for their confidence in the idea and subject matter of the collection. We are also highly appreciative of Rohini Krishnan at Newgen Imaging Systems in Chennai, India, for her role in the production process.

Aziz is particularly grateful for the support of Gada Mahrouse, Eric Shragge, Steve Jordan, Chris Rahim, Radha D'Souza, Leigh Cookson, Cherryl Smith, and Sunera Thobani, as well as the many activists and community organizers with whom he has worked over many years, and for countless conversations, discussions, rants, arguments, emails, debates, inspiration, laughter, and hope. Those contributions—and the social struggles he has been part of—have extended his own understandings and political engagement and nourished this project.

Dip thanks the Dalit and Adivasi communities of South Orissa, the members of the ADEA, VICALP, and the Center for Research and Development Solidarity (CRDS) for their trust and friendship over a decade of partnership and solidarity, and for the inspiration to continue to bring together knowledge contributions (as in this collection) that foreground people's knowledge and learning through struggle and praxis to address injustice. These engagements in turn are made possible with the love and support of family, and their generosity of spirit and continued support will always be treasured.

CHAPTER 1

Learning from the Ground Up: Global Perspectives on Social Movements and Knowledge Production

Aziz Choudry and Dip Kapoor

The dynamics, politics, and richness of knowledge production within social movements and activist contexts are often overlooked in scholarly literature, and sometimes even in the movements themselves. Given the academic emphasis on whether an action, campaign, or movement can be judged a "success," the intellectual work that takes place in movements frequently goes unseen, as do the politics, processes, sites, and locations of knowledge production and learning in activist settings. Even social movement scholarship that draws upon or is embedded in movement actor perspectives has an expressed interest in "taking the measure of the new movements" (see Tom Mertes, 2004, p. xi, a collection of interviews with activists, originally published in *New Left Review*). The contributors to this collection, however, suggest that many powerful critiques and understandings of dominant ideologies and power structures, visions of social change, and the politics of domination and resistance in general emerge from these spaces and subsequently emphasize the significance of the knowledge-production dimensions of movement activism. *Learning from the Ground Up* also challenges ways in which grassroots and movement voices are often overwritten or otherwise marginalized in the context of purportedly "alternative" civil society networks and nongovernmental organizations

(NGOs). The interdisciplinary approaches adopted by the authors in this volume are as rich as the varied movements, processes, and dynamics of knowledge production that these chapters explore and elucidate.

The collection brings together significant and diverse movement actors and sites of knowledge production from both the global South and the global North, engaging in and elaborating upon understandings that are rooted in the contexts of recent or contemporary social struggles, and exploring their knowledge, learning, and educational dimensions. The authors are located within, and/or are working closely with a range of anticolonial, anti-imperialist movements, communities, and popular mobilizations in varied contexts in Asia, Africa, Europe, the Americas, and the Pacific. The collection sets out to articulate and document knowledge production, informal learning, and education work that occurs in the everyday worlds of social activism, highlighting interconnections/dialectics between such knowledge and praxis/action, while illustrating tensions over whose knowledge and voice(s) are heard. This is not intended to be a global or authoritative overview of the state of the world's social movements and knowledge production within movements, although the contributions here highlight several dynamics and tensions that will be familiar to diverse constituencies and contexts.

As we argue elsewhere (Choudry, 2007, 2008; Kapoor, 2009a), the voices, ideas, perspectives and theories produced by those engaged in social struggles are often ignored, rendered invisible, or overwritten with accounts by professionalized or academic experts. In the realm of academic knowledge production, original, single authorship is valued, which inadvertently contributes to a tendency to fail to acknowledge the intellectual contributions of activism, or to recognize the lineages of ideas and theories that have been forged outside of academe, often incrementally, collectively, and informally. That said, we do not intend to imply that these various epistemologies of knowledge (academic and activist) and processes of knowledge production and learning (formal, nonformal, and informal) necessarily exist in completely separate universes.

A major thread of argument in this book emerges from our sense that many scholarly, NGO, and activist accounts pay inadequate attention to the significance of low-key, long-haul political education and community organizing work, which goes on underneath the radar, as it were; that is, even recent and needed critical scholarship (see Philip McMichael, 2010) pertaining to local "critical struggles for social change" remain mostly "development-centric" and subsequently overlook vital "movement-centric" knowledge and learning dynamics. Also in relation to academic scholarship, scholars such as Walter Mignolo (2000) and Maori educationalist Linda

Tuhiwai Smith (1999) raise important questions about the relationship between academic research, epistemologies of knowledge, colonialism, and imperialism, and caution against uncritically applying and overextending theories and concepts developed in Western contexts to third world and Indigenous communities, for example. It is therefore incumbent upon movement and scholar actors to pay closer attention to the specifics of knowledge and learning emergent in and from particular movement contexts in order to avoid these tendencies and to redirect interpretations and related praxis in the interests of the integrity of each movement's politics.

Griff Foley's (1999) *Learning in Social Action: A Contribution to Understanding Informal Education* makes a vital contribution to theorizing the incidental learning processes arising from and contributing to engagement in a range of social struggles. Foley emphasizes the importance of "developing an understanding of learning in popular struggle" (p. 140). His attention to documenting, making explicit, and valuing incidental forms of learning and knowledge production in social action is in keeping with others who understand that critical consciousness, rigorous research, and theory can and do emerge from engagement in action and organizing contexts, rather than as ideas developed elsewhere by movement elites and dropped down from "above" to "the people" (Bevington & Dixon, 2005; Kelley, 2002; Kinsman, 2006; Smith, 1999). In doing so, Foley cautions that although learning through involvement in social struggles can indeed transform power relations, it can also be contradictory and ambiguous, and indeed can sometimes support the status quo.

Academic scholarship that seeks to understand social movement and NGO networks and the education, learning, and knowledge production associated with them must attend to questions emerging from social movements and activist research with respect to relations of power and the ways in which certain forms of knowledge are valued over others. These questions are often based on sophisticated macro–micro analyses of what, to an outsider, might seem a baffling network of relations and shifting power dynamics. We are not claiming here that all learning, evaluation, and analysis embedded in various forms of activism are always *necessarily* rigorous or adequate. Political activist ethnographers George Smith (2006) and Gary Kinsman (2006) urge that activist researchers must go beyond "commonsense theorizing" and attend to actual social practices and organization in the course of social justice struggles. Reflexivity is crucial when starting from, engaging with, and analyzing activist knowledge. Similarly, for Foley (1999) the process of critical learning involves people in theorizing their experience: They stand back from it and reorder it, using concepts like power, conflict, structure, values, and choice. He also emphasizes that critical learning is gained informally, through experience, by acting and reflecting on action, rather than in formal courses.

In his work on social movements and radical adult education, John Holst (2002) refers to the "pedagogy of mobilization" to describe the learning inherent in the building and maintaining of a social movement and its organizations.

> Through participation in a social movement, people learn numerous skills and ways of thinking analytically and strategically as they struggle to understand their movement in motion.... Moreover, as coalitions are formed people's understanding of the interconnectedness of relations within a social totality become increasingly sophisticated. (Holst, 2002, pp. 87–88)

In their thoughtful article on "movement-relevant theory," Doug Bevington and Chris Dixon (2005) note that just as few activists read social movement theory, important debates inside movement networks often do not enter the literature about social movements. They contend that social movement scholars do not have a monopoly on theory about movements. They call for recognition of existing movement-generated theory and of dynamic reciprocal engagement by theorists and movement activists in formulating, producing, refining, and applying research. "Movement participants produce theory as well, although much of it may not be recognizable to conventional social movement studies. This kind of theory both ranges and traverses through multiple levels of abstraction, from everyday organizing to broad analysis" (p. 195). Robin Kelley (2002) concurs with Bevington and Dixon that social movements generate new knowledge, questions, and theory, and emphasizes the need for concrete and critical engagement with the movements confronting the problems of oppressed peoples. He argues that "too often, our standards for evaluating social movements pivot around whether or not they 'succeeded' in realizing their visions rather than on their merits or power of the visions themselves" (p. ix). Kelley reminds us of the importance of drawing conceptual resources for contemporary struggles from critical readings of histories of older movements. This theme runs through most of the chapters in this collection as contributors, in various ways, contextualize their discussion, or otherwise draw upon earlier movements or previous phases of struggles.

In the considerable body of scholarly literature on adult education and learning, thus far relatively few attempts have been made to theorize informal learning and knowledge production through involvement in social action. Foley's (1999) validation of the importance of incidental learning taking place in a variety of social struggles suggests that to do such analyses in order to flush out these dimensions, "one needs to write case studies of learning in struggle, making explanatory connections between the broad political and economic context, micro-politics, ideologies, discourses and learning" (p. 132). This collection is partially motivated by a need to provide a response to such challenges to better understand knowledge production and learning in social movements and activism.

As coeditors of this volume, our own location informs and is informed by our various engagements in struggles for social change. Choudry's interest in this topic comes from more than two decades of social and environmental justice organizing, popular education, research, and writing in activist groups, social movements, and NGO networks and coalitions in Aotearoa/New Zealand, the Asia-Pacific region, Canada, and globally. Many of his scholarly interests arise from a sense of disjuncture shared with many others engaged in social struggles around the gulf between their experience and knowledge, and what gets written, thought, and taught about social justice, social movements, community organizing, and activism in scholarly circles—questions of who writes, for whom, and for what purpose, and implicit and explicit hierarchies of knowledge production and education. A board member of the Immigrant Workers Centre in Montreal, he is also a co-initiator and member of the editorial team of the multilingual, collaborative Web site www.bilaterals.org, which was established by several NGOs and activist groups to support critical analysis of and resistance against bilateral free trade and investment agreements. He has long been involved in support for self-determination struggles of Indigenous Peoples, and an anticolonial framework of analysis informs both his activist and academic practice.

Kapoor's work in Orissa, India, as a popular educator and researcher, organizer, and participant in Adivasi (original dweller) and Dalit (downtrodden caste groups pejoratively referred to as untouchables) existential struggles (political-economic and cultural) since the early 1990s informs his political and scholarly efforts and engagements pertaining to Adivasi/Dalit movement knowledge/learning interests and social change engendered by caste-colonial histories and contemporary dispossessions being produced by the globalization of capitalism (for instance, see Kapoor, 2009b). As cofounder and research associate for the Adivasi-Dalit people's Center for Research and Development Solidarity, popular research and activism collude to stimulate movement knowledge production and sociopolitical impact in keeping with Adivasi-Dalit political projects and movement teleology. Current research projects, for instance, include an analysis and activation (simultaneous engagements) of learning in Adivasi movements; Adivasi-Dalit unity and state-corporate-caste interests and disunity; knowledge politics in NGO-movement relations; movement-to-movement relations and prospects for translocal activism; leaders' and movement histories for movement building; and movement strategy and mining/industrial incursions.

Challenging the Script: The Importance of Documentation and History

Part of the challenge of putting this book together relates to the pressures on many of the contributors to find time away from the everyday world of organizing to write and reflect on the themes of this book. As Rahila Gupta (2004) of Southall Black Sisters in the UK notes, it is not easy for activists "to sit down and record their work, but in this age of information overload you need to record in order almost to prove that you exist" (p. 3). As many of the contributors in this volume explicitly state (for example, Rodriguez, Choudry, Desai and Walsh, Ziadah and Hanieh), it *is* important to document the articulation of challenges to hegemonic NGO and "civil society" positions, otherwise these will prevail unchallenged as the definitive "alternative" discourses to be referenced by, and inform, future movements and academic inquiry.

As editors, and as is apparent from the collection as a whole, we are also cognizant of the importance of the contribution of critical approaches to history in understanding the relationship between social movements, knowledge production, learning, and action. At the same time as we suggest that they can be rich sources of ideas for social transformation, NGO, "alternative," and movement accounts should also be subjected to critical scrutiny and not romanticized. Indeed, some speak about the emergence of a nonprofit industrial complex (INCITE! Women of Color Against Violence, 2007), which undermines people's struggles and exalts professionalized NGO careerism and corporate models of organization over democratic, mass-based organizing. Further, as a number of chapters in this volume contend, professionalized NGO or academic writers often author these accounts, and in some ways replicate the tendency that Eschle (2001), Sarkar (1983 and 1998), and others charge academic scholarship with; that is, ignoring or misconstruing forms of social action that do not fit within a preestablished theoretical framework. There is a danger that the conventional processes of production of academic scholarship, and assumptions or claims that such activity constitutes the apex of intellectual rigor and inquiry, can in fact overlook the complexities and dynamics of activism, and the intellectual contributions of activist practice (Bevington & Dixon, 2005; Frampton et al., 2006).

Studies of knowledge production and mobilization in activist, trade union, and NGO networks must attend to their specific geohistorical context and the actual social forces in which they are implicated, going beyond objectifying kinds of analyses. We also need to interrogate the silences, omissions, and misrepresentations that may exist in dominant NGO and scholarly accounts of campaigns and mobilizations. In different ways, the politics of knowledge production and the social movements considered in this book are located in older histories of struggle against colonialism and imperialism. Elsewhere, Mathew (2005), one of the contributors to this volume, argues that in order to understand contemporary social movements in the third world, and their ideas, we have to take history seriously:

> Maybe our collective task... is to look carefully at the resurgent left social movements all across Africa, Asia, and Latin America and comprehend the ideas of justice that inhere within these movements and the historical memory they are rooted in. (p. 203)

The traditions, trajectories, hopes, visions, and dilemmas of past and present struggles not only are rich sources for extending academic scholarship, but also offer vital tools for contemporary activism.

Organization of the Book

The book is organized in three parts, although we recognize that there are many intersections, cross-cutting themes, and ways in which the contributions are in constant dialogue with each other. Part I discusses tensions, challenges, and dilemmas concerning knowledge and learning in various contexts of politics of knowledge production, representation, and relationships between "civil society," academe, and social activism.

Choudry critically interrogates processes and practices of "NGOization" of social change in the "global justice movement." Drawing from his involvement in anti-APEC (Asia-Pacific Economic Cooperation) and anti-WTO (World Trade Organization) activism, he argues that a relatively small NGO elite attempts to claim positional superiority for forms of professionalized knowledge and advocacy that effectively sideline, filter, or erase more critical positions opposed to capitalism and colonialism. He contends that the dominant tendency of NGOs to compartmentalize the world into "issues," their unwillingness to name or confront capitalism, and an NGO "ideology of pragmatism" serves to undermine and contain more critical forms of knowledge production and action in relation to confronting global capitalism, but suggests that this is also being challenged by ideas and mobilization strategies arising from past and present anti-imperialist and anticolonial struggles.

Desai and Walsh discuss knowledge production, power, and struggle in relation to resistance to xenophobic violence against migrant workers and refugees from other African countries in South Africa since the end of apartheid, taking issue with the erasure of migrants' perspectives and accounts accomplished through dominant forms of framing employed by state institutions, academic researchers, and NGOs alike. They examine the politics of knowledge production that silenced the voices and illustrate how firsthand accounts of refugees challenge these accounts.

Rodriguez examines "civil society" processes around the 2008 intergovernmental Global Forum on Migration and Development (GFMD) in Manila. Grassroots migrant activists countered the government meetings, but perhaps more important, explicitly countered officially sanctioned "civil society" meetings with their own international assembly where they declared that they would "speak for themselves." Her chapter examines the sets of knowledge mobilized and deployed by migrants themselves as compared with those of their so-called civil society "advocates." She asks: How are notions of "expertise" and "authority" defined by different actors, and to what political ends are particular forms of expertise, authority, and knowledge used?

Hudig and Dowling draw from their involvement in European counter-globalization movements and commitment to principles of horizontal self-organized networks with particular attention to the European Social Forum processes and summit protests to consider questions of vanguardism, professionalization, and knowledge production. Arguing that this movement has reached an impasse, they ask what can be learned from struggles over modes and dynamics of organizing in Europe in recent years. Viewing knowledge production as an integral part of the political practice of social movement activism taking its starting point from concrete experiences in thinking through the frustrations encountered, they formulate these as sites of inquiry that can reveal insights into possible steps forward. In doing so, they discuss the vulnerability of horizontal self-organized networks, and the contentious role of NGOs and vanguardist political parties in these mobilizations.

Ziadah and Hanieh contribute a chapter on collective approaches to activist knowledge in their experiences of Palestine solidarity organizing in Toronto, Canada, a city that has seen many successes for Palestine solidarity work over the past few years. They reflect on major lessons of Palestine solidarity movement building in Toronto over the past five years. This discussion is framed around the importance of maintaining an open, self-sustaining culture of knowledge production and skills transfer within this activist work. They contend that because of its fundamentally democratic and collective understanding of the ownership of knowledge and skills, this approach runs up against the socialization that individuals face within a capitalist society. They analyze the types of structures and organizing principles employed by Toronto-based movements in order to facilitate this knowledge-building approach within activist work. This includes issues such as nonsectarianism, multigenerational organizing spaces, democratic and clear structures, and a radical but open politics. They conclude by reflecting on what this means for rebuilding and strengthening broader networks on the left.

Hsia draws from her direct participation in the empowerment of the "foreign brides" and the making of an im/migrant movement in Taiwan, highlighting the role of knowledge that has been produced from long-term praxis in these struggles. Drawing from Touraine (1988), she discusses the "subjectivation" process of these marriage migrants—mainly women from Southeast Asia who have married Taiwanese men—as they develop their sense of individual and collective agency, in part through Chinese language lessons, and the formation of their own activist organization with the support of Taiwanese feminist activists. She discusses the positions of intellectuals in social movements, arguing that intellectuals should see themselves as the "conscientious wolfman" rather than the leaders in the movements. By using this metaphor, the author argues that with more access to resources, intellectuals involved in movements should devote themselves to the empowerment of the masses, avoiding the possibilities of wounding and betraying the movements in the future.

Part II discusses making knowledge and learning from trade unions, worker alliances, and left party-political activism. Novelli explores trade union resistance to privatization through a case study of SINTRAEMCALI, a public service union located in southwest Colombia that has successfully fought off attempts by the national government to privatize public utilities in the country's second city of Cali. The case study charts the emergence of a strategic pedagogy within SINTRAEMCALI that linked workers and local communities in the defense of public services and operated on a range of scales from the local to the global. These pedagogical spaces were zones of resistance where alternative ways of knowing, learning, and acting were developed and fought over and new strategies were produced. Novelli examines how the process forged a network of solidarity links that transcended local and global boundaries and linked up a range of sites of resistance containing differential capacities and resources that the trade union could access and deploy in the struggle to defend public services.

Bleakney and Morrill discuss struggles, pressures, and challenges facing contemporary trade unions and labor education in Canada and the United States. Drawing upon their combined experience as labor educators and activists, they discuss external and internal tensions of trade unions and union education in an era of global neoliberalism. In highlighting the important contributions that a social-movement-oriented union education can make to workers and broader movements for social change, they also discuss some of the strategies, struggles, and methods that have been used to push labor education on both sides of the border to move beyond "traditional" and "banking" models.

Mathew discusses the nature of knowledge production in the context of political mobilization of the New York Taxi Workers Alliance, arguing that this process is core to continuing political work and the development of theory. Drawing from engagement in and reflection on a recent two-year period of mobilization by taxi drivers against the imposition of GPS technologies in a campaign that led to a series of strikes in 2007, the chapter also considers the role of organizers and intellectuals in such political mobilizations. Mathew discusses the processes and significance of the development of knowledge around the consequences of this technology. He argues that this knowledge was not only crucial for the shape that the campaign took, but it also broke down divisions between the organizers and intellectuals as knowledge producers and the drivers (workers/mass base) as knowledge consumers. The chapter contends that the process of abstraction that is constitutive of theory produces a fundamental division between worker and organizer, despite the critical role that workers have played in the production of knowledge, and signals one possible reason for the demise of new cycles of dialectical growth in organizing work, unless the workers' organization actively pays attention to the creation of intellectuals within the movement.

Austin explores what can be learned from the short-lived (1979–1983) Grenada Revolution, arguing that this revolution embodied in microcosm many of the challenges that confronted twentieth-century socialism and social movements, and that Grenada's experience has profound and universal implications for our understanding of social transformation and the dynamics of liberation. He explores a number of ideological and geopolitical factors that led to the implosion of the revolution, arguing that the revolution's leadership's messianic faith in Marxist-Leninist theory and blind dependence on vanguardist politics distanced it from the majority of Grenadians. Lessons from the Grenada Revolution, says Austin, are not only important to the revival of the Caribbean left; the world, particularly the generic left, also has much to learn from Grenada's experience.

Part III focuses on the production of knowledge and learning in and from peasant and Indigenous struggles. Prasant and Kapoor draw from a decade-long experience of organized activism by the Adivasi-Dalit Ekta Abhijan

(ADEA) rural movement in South Orissa, India, to address Dalit concerns pertaining to land and forest alienation and the perpetuation of poverty among Dalits. Their chapter considers ADEA critical-political analysis with respect to Dalit marginalization and the *politics of division* cultivated through (a) divisive legislation and land and forest classification schemes, and (b) through fomenting caste-communalism (Hindutva-inspired violence) and resulting forced migrations of Dalits to enable upper caste-state-corporate industrial land grabs for the development of Special Economic Zones (SEZs) in the contemporary Indian neoliberal political-economy. Key learning and knowledge produced through movement activism (and the significance of the same for the movement) to address the politics of caste-class domination and resistance are shared, along with some concluding reflections on the prospects for Dalit-Adivasi (original dweller/tribal) rural activism.

Across the border in Pakistan, Sayeed and Haider consider knowledge, power, and struggle in the context of resistance by sharecroppers working on Pakistan military farms and the Pakistan Seed Corporation as they contest their right to land and livelihood. Organized under the umbrella of Anjuman-e-Mazareen Punjab (AMP), and the slogan *"Maliki ya Maut"* (Urdu: "Ownership or Death"), this movement remains active after a decade of intense state repression. The chapter explores the various strategies employed by the AMP to retain control over the land, internal tensions within the movement, and the methods employed by various political, economic, and military elites to maintain control over land even in the face of the widening split that the movement faces.

Pictou and Stiegman discuss the relationship between learning in action, education, and resistance to colonial capitalist state impositions of commercial models of fisheries on the Mi'kmaq people of Bear River First Nation (BRFN) in Nova Scotia, Canada. They argue that for BRFN, it is impossible to express spiritual and cultural values through a fisheries management regime predicated on resource privatization, individual property rights, and corporate profits, and that is hostile to the contributions of Mi'kmaq traditional knowledge. Their chapter highlights the community's experience of learning through social action over the past decade. It discusses internal struggles to articulate a vision for authentic Mi'kmaq "development," as well as the role of knowledge, learning, and community media in building alliances with local nonindigenous fishing groups resisting privatization of fisheries. It also describes the impacts of methodology and the methodology used for collaborative filmmaking on these issues between a nonindigenous graduate student (Stiegman), a First Nations community, and neighboring nonindigenous fishers.

References

Bevington, D., & C. Dixon. (2005). Movement-relevant theory: Rethinking social movement scholarship and activism. *Social Movement Studies* 4(3), 185–208.
Choudry, A. (2007). Transnational coalition politics and the de/colonization of pedagogies of mobilization: Learning from indigenous movement articulations against neo-liberalism. *International Education* 37(1), 97–112.
Choudry, A. (2008). NGOs, social movements and anti-APEC activism: A study in power, knowledge and struggle. Unpublished PhD dissertation, Concordia University, Montreal.
Eschle, C. (2001). Globalizing civil society? Social movements and the challenge of global politics from below. In *Globalization and social movements*, ed. P. Hamel, H. Lustiger-Thaler, J. Nederveen Pieterse, & S. Roseneil, pp. 61–85. Basingstoke: Palgrave.
Foley, G. (1999). *Learning in social action: A contribution to understanding informal education*. London & New York: Zed Books.
Frampton, C., G. Kinsman, A. K. Thompson, & K. Tilleczek. (2006). Social movements/social research: Towards political activist ethnography. In *Sociology for changing the world: Social movements/social research*, ed. C. Frampton, G. Kinsman, A. K. Thompson, & K. Tilleczek, pp. 1–17. Black Point, NS: Fernwood.
Gupta, R. (2004). Some recurring themes: Southall Black Sisters 1979–2003 and still going strong. In *From homebreakers to jailbreakers: Southall Black Sisters*, ed. R. Gupta, pp. 1–27. London: Zed Books.
Holst, J. D. (2002). *Social movements, civil society, and radical adult education*. Westport, CT: Bergin and Garvey.
INCITE! Women of Color Against Violence (eds.) (2007). *The revolution will not be funded: Beyond the non-profit industrial complex*. Boston, MA: South End Press.
Kapoor, D. (2009a). Participatory academic research (par) and People's Participatory Action Research (PAR): Research, politicization and subaltern social movements (SSMs) in India. In *Education, participatory action research, and social change*, ed. D. Kapoor & S. Jordan, pp. 29–44. New York: Palgrave Macmillan.
———. (2009b). *Adivasis* (original dwellers) "in the way of" state-corporate development: Development dispossession and learning in social action for land and forests in India. *McGill Journal of Education* 44(1), 55–78.
Kelley, R.D.G. (2002). *Freedom dreams: The black radical imagination*. New York: Beacon Press.
Kinsman, G. (2006). Mapping social relations of struggle: Activism, ethnography, social organization. In *Sociology for changing the world: Social movements/social research*, ed. C. Frampton, G. Kinsman, A. K. Thompson, & K. Tilleczek, pp. 133–156. Black Point, NS: Fernwood.
Mathew, B. (2005). *Taxi! Cabs and capitalism in New York City*. New York: New Press.
McMichael, P. (ed.) (2010). *Contesting development: Critical struggles for social change*. New York: Routledge.

Mertes, T. (2004). *A movement of movements: Is another world really possible?* London: Verso.

Mignolo, W. D. (2000). *Local histories, global designs: Coloniality, subaltern knowledges and border thinking.* Princeton, NJ: Princeton University Press.

Sarkar, S. (1983). *Popular movements and middle class leadership in late colonial India: Perspectives and problems of a "history from below."* Calcutta: Centre for Studies in Social Sciences.

———. (1998). *Writing social history.* Delhi: Oxford University Press.

Smith, G. W. (2006). Political activist as ethnographer. In *Sociology for changing the world: Social movements/social research*, ed. G. Kinsman, A. K. Thompson, & K. Tilleczek, pp. 44–70. Black Point, NS: Fernwood.

Smith, L. T. (1999). *Decolonising methodologies: Research and indigenous peoples.* Dunedin: University of Otago Press and London: Zed Books.

Touraine, A. (1988). *Return of the actor.* Minneapolis: University of Minnesota Press.

PART I

Making Knowledge and Learning from the Politics of Knowledge Production and Representation: "Civil Society," Academe, and Social Activism—Tensions, Challenges, and Dilemmas

CHAPTER 2

Global Justice? Contesting NGOization: Knowledge Politics and Containment in Antiglobalization Networks

Aziz Choudry

Introduction

> More insidious than the raw structural constraints exerted by the foundation/state/non-profit nexus is the way in which this new industry grounds an epistemology—literally a *way of knowing* social change and resistance praxis—that is difficult to escape or rupture.... [T]he non-profit industrial complex has facilitated a bureaucratized *management of fear* that mitigates against the radical break with owning-class capital (read: foundation support) and hegemonic common sense (read: law and order) that might otherwise be posited as the necessary precondition for generating counter-hegemonic struggles. (Dylan Rodriguez, 2007, p. 31)

This chapter discusses struggles over power and knowledge among nongovernmental organizations (NGOs) and social movements contesting capitalist globalization. Building on a review of an emerging body of critical literature on NGOs' implications in capitalist relations and the professionalization of social change, and the author's activist engagement, it argues that processes of NGOization and professionalization, and hierarchies of power and knowledge within "alternative" milieus often reproduce rather than challenge dominant practices and power relations, and serve elite economic and political interests instead of constituencies which these organizations claim to represent. It identifies and questions aspects of hegemonic NGO practices, primarily among "civil society" networks contesting the Asia-Pacific Economic Cooperation (APEC)[1] forum in the 1990s, in which I was an activist. In interrogating "alternatives to globalization" advocacy positions advanced by nongovernmental actors ostensibly committed to social transformation, I argue that critical attention to texts and actual practices reveals that many NGOs are tied to what I refer to as an ideology of pragmatism which normalizes and reinforces dominant ideologies of liberalism and liberal democratic nation-states with regard to their current and historical implication in colonialism and global capitalist relations. In contending that many such "alternative" actors claim a positional superiority for a small, professionalized NGO/activist elite, this chapter highlights how NGO texts and actual practices can socially organize, conceptually coordinate, fragment and compartmentalize struggles for social and environmental justice, undermining or constraining more critical systemic analysis. The concluding section discusses some challenges to these trends arising from knowledge, power, and movement activism from below.

NGOs, Neoliberalism, Pragmatism, and Compartmentalization

During the 1990s, NGOs and "civil society" organizations and rhetoric mushroomed exponentially worldwide, alongside an expansion in the formal or de facto subcontracting of NGOs by states and international financial institutions. Increasingly, governments, intergovernmental organizations, and international financial institutions promoted "strengthening civil society" and "good governance"—both intrinsic pillars of a neoliberal policy environment. The dominant notion of "civil society" emphasizes the rights of individuals to pursue their self-interest rather than collective rights, and simultaneously upholds and obscures the interests of state and capital. It facilitates what Kamat (2004) calls the privatization of the notion of public interest. Wood (1995) warns:

> "Civil society" has given private property and its possessors a command over people and their daily lives, a power enforced by the state but accountable to no one, which many an old tyrannical state would have envied... [T]he cult of civil society also tends to reproduce the mystifications of liberalism, disguising the coercions of civil society and obscuring the ways in which... state oppression itself is rooted in the exploitative and coercive relations of civil society. (pp. 254–256)

As the triumphalism of "civil society" spread throughout the last two decades, the term "NGOism" has become increasingly widely used in social movement activist networks. Many movement activists warn of the "NGOization" of movements and struggles; that is, their institutionalization, professionalization, depoliticization, and demobilization (Armstrong & Prashad, 2005; Kamat, 2004; A. Smith, 2007). Kamat (2004) argues that this process is driven by the neoliberal policy context in which NGOs operate. Organizations must demonstrate managerial and technical capabilities to administer, monitor, and account for project funding. Mass-based organizations of movements that assert themselves through various forms of political mobilization are often displaced by or are in considerable tension with organizations that claim to represent the poor and marginalized, but which have no mass base or popular mandate and less critical platforms (McNally, 2002; Petras & Veltmeyer, 2001, 2003, 2005). Piven and Cloward (1977) argue that the preoccupation with financial survival and building and maintaining organizations diverts energy and resources away from organizing and escalating popular protest movements, and indeed often blunts or curbs them.

In NGO networks, there is much focus on development and development models, which often obscures the capitalist assumptions which underpin them. Frequently, these organizations merely seek to ameliorate some of the social or environmental impacts through community development and participation-based development projects. Wood (1995) succinctly describes some intellectual and conceptual dilemmas that help to explicate some of the cleavages of NGO positions regarding neoliberalism, and the implications for the NGOization of political space in general:

> At the very moment when a critical understanding of the capitalist system is most urgently needed, large sections of the intellectual left, instead of developing, enriching and refining the required conceptual instruments, show every sign of discarding them altogether.... Intellectuals on the left, then, have been trying to define new ways, other than contestation of relating to capitalism. The typical mode, at best, is to seek out the interstices of capitalism, to make space within it for alternative "discourses", activities and identities. (pp. 1–2)

McNally (2002) argues that for the 'antiglobalization movement', failure to name capitalism comes at great cost, encouraging the movement's supporters to see the problem not as the system that organizes our lives, but merely as a set of policies pursued by those currently at the top in need of reform. The effect is to deradicalize the movement by proposing merely to change the ideology that drives domestic government policy and intergovernmental decision-making, not the system as a whole. Consistent with this, critics reluctant to name capitalism as the problem typically advocate building a citizen's lobby to urge reforms within a capitalist framework.

This frequent failure to name capitalism, imperialism and colonialism, alongside commonly articulated NGO platforms on participation, fair trade, sustainable development, vaguely defined claims about democracy, rights and justice, and sometimes a kind of stylized, "respectable" militancy, also helps to obfuscate the ways in which many of these organizations are implicated in ruling relations, forms of discursive and practical organization which coordinate these activities and actors in the interests of the state and capital (G. W. Smith, 2006). All too often, NGOs have defined themselves solely in terms of fitting into the existing structures of power to supposedly represent the interests of the people or "civil society".

Indeed, with the ascendancy of NGOs and the enlarging of the political space that they took up arose forms of hegemonic NGO politics. For Kamat (2004), rather than "deepening the gains made on the basis of popular democratic struggles, NGOs are being re-inscribed in the current policy discourse in ways that strengthen liberalism and undermine democracy" (p. 171).

As Petras and Veltmeyer (2001), McNally (2002), Kamat (2004), A. Smith (2007), and others charge, many of these organizations explicitly orient their focus upward toward the structures of power, seeking legitimacy through lobbying and reporting and maintaining relations with state institutions and other national and international funders, rather than through building or resourcing a mass movement base with a capacity to meaningfully challenge existing power relations. Moreover, some charge that even though some NGOs have gained access to policy-makers in relation to trade policy, this has largely failed to bring about substantive changes in policy outcomes—"inclusion without influence," as Dür and De Bièvre (2007, p. 86) put it.

Many of these organizations adhere to an *ideology of pragmatism* that assumes the most that can be hoped for in terms of social change are limited gains as opportunities permit *within* existing structures. Their praxis and principles usually rooted in liberal notions about society and the state, many NGOs have vehemently denounced or stigmatized activists and movements that drew from Marxist and other explicitly anticolonial or anti-imperialist traditions as doctrinaire, anachronistic ideologues, sometimes excluding them from their activities, at other times shutting down more critical voices in NGO events, amply illustrating Rodriguez' (2007) concept of a "bureaucratized management of fear" (p. 31). This framing invites us to accept a false binary that some movement or NGO actors are "ideological" and others are apolitical or non-partisan. In their professionalized and institutionalized realities and daily activities, and given the ways in which many NGOs are socially organized in the interests of ruling through funding, regulation, and patronage from elites, they enact what is often a doctrinaire ideology *of* pragmatism. They operate in spaces within what Wood (1995) calls the "interstices of capitalism" rather than seeking to transform the system. Their discourses may sound invitingly progressive at face value, but they are often disconnected from social movements that confront state and capital. Instead many of these organizations focus on lobbying and trying to influence elites, more driven by notions of polite reformism and self-interest in the maintenance of their organization and funding relationships—and ultimately elite interests (McNally, 2002; A. Smith, 2007). In some cases, these organizations have become corporate entities in their own right, part of what Rodriguez (2007) and Andrea Smith (2007) describe as "the non-profit industrial complex" (A. Smith, p. 3), modeled after capitalist structures.

Professionalized Intellectual Policemen? NGOs, Organizing Knowledge, and Gatekeeping

Petras and Veltmeyer (2001) view the vast majority of NGOs as "intellectual policemen who define "acceptable" research, distribute research funds, and filter out topics and perspectives that project a class analysis and struggle perspective" (p. 137). While we should be wary of applying a totalizing framework of analysis to understand NGOs, many serve as gatekeepers, regulating access to knowledge about struggles in the South, and indeed local mobilizations for social and economic justice. Clifford Bob (2005) suggests that "In their role as gatekeepers, major NGOs may act as brakes on more radical and exceptional ideas emanating from the developing world, and for that reason some important challengers eschew foreign ties" (p. 194). Some NGOs—especially aid and development agencies with funding relationships with organizations in the South, and some research and advocacy NGOs in both the North and South—position themselves as gatekeepers between social movements and organizations in different parts of the South. That is, they act as intermediaries, and yet their roles and interests in doing so, and the power inherent in acting in this way are frequently invisible and rarely subject to critical examination. Townsend and Townsend (2004) note that gatekeeper NGOs "command the discourse, can write the funding proposals...and are 'in the information loop'" (p. 281), often creating a sense of powerlessness for those on the outside. Northern NGOs and social movement activists sometimes seem often unaware of or unconcerned whether these Southern organizations have a genuine grassroots base, or, rather, represent a professional class of NGO representatives with access to international networks. In their eagerness to demonstrate third-world connections, these NGOs tend to link internationally with people most like them—NGO professionals whose practice and discourse is rooted in liberal traditions.

"Global justice" networks are often uneasy, usually loose coalitions, and many of the organizations involved approach trade and investment liberalization through a compartmentalizing lens. Armstrong and Prashad (2005) contend that coalitional politics has positives and negatives. They see it as a result of fragmentation and the "NGOization" of the left. "[E]ach of our groups carves out areas of expertise or special interest, gets intensely informed about the area, and then uses this market specialization to attract members and funds. Organizations that 'do too much' bewilder the landscape" (p. 184). While they argue that specialization can result in valuable analytical and strategic resources for a broader movement, the authors suggest that fragmentation is problematic because it leaves us without a sense of common strategy, tactics, or movement, or political agreement about how the systems currently operate and reproduce themselves. They write:

> As we extend coalitional politics from our local and national contexts to webs of networking fostered by international conferences, we need to ask again about how we know what we fight, and what alternate futures we see emerging from our often delinked, but not disparate struggles. (pp. 184–185)

In the absence of a unifying vision or platform, these questions are highly relevant for transnational coalitions and networks. One significant way in which many NGOs organize knowledge is the tendency to compartmentalize issues and struggles, which replicates the dominant systems of knowledge based on what Linda Smith (1999) calls a colonially driven "systematic fragmentation" (p. 28), disciplinary carve-up and disconnection of peoples from their histories, landscapes, social relations, and ways of thinking, feeling, and interacting with the world. The process of systematic fragmentation that she describes has arguably influenced "civil society" networks in terms of which voices are heard and valued, and which movements and organizations are seen as representative and accountable to a broader base. The strong tendency of NGOs toward a project-by-project approach reinforces this compartmentalization and obscuring of the overarching framework of imperialism in which we struggle.

For Biel (2000), NGOs as key actors in a liberal pluralist civil society are central to a "new political economy of co-opted empowerment," which promotes fragmentation and inhibits "the gathering-together of the forces of the poor" (p. 298). Alongside this, the exaltation and reification of a class of "civil society" experts within "alternative" networks, that is not grounded in a struggle perspective is often valued over a more broadly framed and more critical/confrontational analysis from social movement organizers (also see Robyn Rodriguez, in this volume, for a discussion on the role of experts in relation to social struggles). At the same time, we should recognize that an awareness of the comprehensiveness and systemic nature of global capitalism's impacts can explain the resistance from broad fronts of social movements in many countries that *do* mobilize and resist against a common enemy. Foley's (1999) work on learning in social action and the emerging field of political activist ethnography (G. W. Smith, 2006) highlight and value the incremental building of knowledge, which arises from actual engagement in our everyday world. Yet Foley and others also highlight the inherently contradictory and contested nature of such knowledge production and forms of social action.

Parallel "civil society"/"people's" forums have shadowed official intergovernmental economic and political summits for some years now, and are themselves key sites for internal struggles over power, knowledge, and representation. During my involvement in transnational anti-APEC activism in the mid- to late-1990s, tensions that became more visible in the post-Seattle/World Trade Organization (WTO) era of global justice activism were prefigured. Alternative APEC NGO conferences and campaigns produced an abundance of documents. Texts and discourse are of great significance to the oppositional movement/NGO networks that contest neoliberal globalization. More rigorous scrutiny of the publications and discourse of large NGOs often assumed to be "on the same side" in global justice campaigns (e.g., Oxfam, 2002) would help reveal the rationale for the way in which these organizations have denounced anticapitalist activists and direct actions (Buckman, 2004), and purport to represent "civil society" in a "dialogue" with governments, international institutions, and other players, while others refuse selective dialogue as a "divide and rule" tactic. Texts are sites of real struggles. For Kinsman (1997), "[d]ocuments can be attempts at providing conceptual organization for the co-ordination of state and professional responses to 'social problems.' They organize knowledge in particular directions and from particular standpoints, which often include the containment of social movements" (p. 216). How do various players such as the media, different publics, governments, NGOs, and more radical groups or movements read and use such texts in particular contexts and moments? Kinsman notes that "[r]egulation is often accomplished through...texts and how they are read and used. They are an important part of the social organization of hegemony, which counterhegemonic politics must address" (p. 216).

NGOs, socially engaged academics, and activists have examined numerous official documents, speeches, and other texts produced to promote global capitalism. Part of this work involves analyzing draft trade and investment agreements and legislation, written in the arcane, technicist language of trade law. These documents redefine and reconceptualize areas of human activity in terms that commodify them and subject them to a marketization discourse. For example, not until the GATT (General Agreement on Tariffs and Trade—now WTO) Uruguay Round negotiations held between 1986 and 1994 were agriculture, investment, services, and intellectual property rights (IPRs) defined as global trade, rather than domestic policy concerns (Kelsey, 1999; Raghavan, 1990). This type of NGO policy analysis and advocacy favors professionalized technical knowledge and seems to value attempts to establish credibility with, and influence upon, governments and the private sector, rather than prioritizing popular education resources that meet the needs of movements engaged in struggles against global capitalism.

Many documents produced by NGOs on APEC and the WTO, for example, reflect a compartmentalized worldview that reproduces the way that free trade and investment agreements redefine broad spheres of human activity in "trade-related" terms. The embrace of neoliberal discourse and capitalist logic in NGO practice serves to tie back analysis and strategy to the interests of capital and to contain or even preclude more radical positions. These NGOs fail to question fundamental assumptions underlying such definitions, remaining instead within the parameters set by international trade negotiations, trade law, and indeed liberal conceptions of the world. For example, dominant NGO concerns about traditional knowledge of Indigenous Peoples or farmers tend to be framed within the concept of IPRs. Advocacy on this issue frequently fails to challenge the fundamental notion and capitalist ideological frame of IPRs, but rather urges reforms of language in particular WTO clauses about access to medicines or the patenting of life-forms, for example (Khor, 2001). Similarly, many NGO discussions of farmers' rights to grow food to sustain their families in the context of agricultural liberalization and domination (Oxfam 2002; Watkins, 1996) do not substantively challenge neoliberal conceptualizations of food and nature as mere tradable commodities, and posit solutions for poor third-world farmers through better market access to rich countries without challenging capitalism (Desmarais, 2007).

For example, in 1996 I spoke on a panel at an "Alternatives to APEC" meeting in the Philippines alongside a senior Oxfam policy adviser who argued for freer trade in agriculture. In his paper, he contended that the problem facing small Southern farmers was comparative access to subsidies and unfair competition with (subsidized) Northern producers; what was needed to address this were better "opportunities for participation in markets" (Watkins, 1996, p. 22), that is, more trade liberalization. For Oxfam, the *rules* of trade may be challenged, but not the fundamentals of the *system* that underpins it. More than a decade later, Oxfam maintains this stance. In turn, social movements such as La Via Campesina (an international peasant and farmers' movement network active in struggles against neoliberal globalization and corporate agriculture), and other NGOs (which Watkins dubbed the "extreme end of the anti-globalization movement", p. 168, Buckman, 2004) have criticized its position for undermining food sovereignty and reproducing dominant positions on globalization.

NGO counter-summits tend to be dominated by organizations from both the North and South that embrace liberal Western ideals and notions of "rights." In APEC parallel forums, many NGO positions considered APEC to lack something (e.g., concern for environmental impacts of economic growth or human rights). Such NGOs would not question APEC's fundamental capitalist logic, but posit that it needed "mending." Some NGOs ineffectively offered themselves as consultants to balance out APEC's agenda, eschewing confrontation. They held that the solution was just a question of augmenting the existing APEC framework through rational dialogue with political and economic elites. These positions reflected a belief that capitalist globalization merely needed some reforms and democratization to deliver environmental and social goals. The closure of the APEC process to any input by NGOs and its fundamentals of free trade and investment clearly made such a position untenable. By contrast, positions that were grounded in popular struggles, and those (usually much smaller) NGOs that supported them, sought to delegitimize APEC and denounced it as a fundamentally antidemocratic process and its economic vision as unreformable. Since most of these more critical movements and organizations focused on building up resistance to neoliberalism locally (while fostering international solidarity links), their energies were directed at change from below, rather than trying to negotiate terms of engagement with governments and the private sector for inclusion in APEC itself. Many of these players openly confronted state power, capital, APEC, and other instruments for capitalist globalization. They strongly advocated strategies of nonengagement with and the delegitimization of APEC, and usually advanced broader systemic critiques of capitalism and colonialism.

Confronting Hegemonic NGO Positions

Countercritiques can be produced from the standpoint of peoples' struggles to challenge and disrupt "NGOism," for example, to expose contradictions, and point to other ways forward. These necessarily need to go beyond responding to official texts and discourse to naming the systems that lie behind economic, social, and environmental injustice. Activists with more critical positions must typically pry open space for different voices to be heard besides those NGOs and "experts" who have come to dominate political space in many countries. The fight back, then, is not only against global capitalism, but also against authoritative "civil society" actors and discourses. In NGO forums this often means trying to raise challenges from the floor, such as the very desirability and effectiveness of the institutionalization of large-scale People's Summits, and demands that participating organizations meaningfully address colonial injustices against Indigenous Peoples, the demands of peasant farmer/landless movements, or militant rank-and-file unionists critical of capitalism and labor bureaucracy, for example. These interventions are often not recorded and are valued far less than written documents or presentations by "experts." Such erasures have consequences for how such mobilizations are understood and what lessons can be learned for future phases of struggle.

Some of the most cogent and systemic challenges to capitalist globalization have emerged from Indigenous Peoples' movements for self-determination, contextualized in longer histories of resistance to colonialism. Elsewhere, I discuss Maori opposition to domestic neoliberal reforms and the imposition of free trade and investment regimes in Aotearoa/New Zealand (Choudry, 2010). I concur with Burgmann and Ure (2004) that the contributions of Indigenous Peoples' struggles for self-determination are very useful for theorizing convincing alternatives. For them, "the practical critique of neoliberalism embodied in indigenous people's resistance to their incorporation into the global market is one informed by an often acute recognition of not only the global dimensions of such resistance but also an acknowledgement of anti-imperialist struggles stretching back over many hundreds of years" (p. 7). This has also "enabled non-indigenous groups and movements to root their critique in an anti-capitalist perspective that emanates from non-Western sources" (p. 8).

In settler-colonial states like Aotearoa/New Zealand, Australia, and Canada, and in the discourse of many NGOs based in these countries, the dominant frame for most "global justice" campaigns typically identifies transnational corporations, powerful governments like the United States, and domestic business and political elites as engines of neoliberalism, but proposes a program of reforms and strengthening social democratic governance as a solution, nationally and internationally. This frame advocates nostalgia for a Keynesian welfare state, retooling the national government, reregulation of the economy, tighter controls on foreign investors, more social spending and more public consultation, and transparency around policy making. Underpinning this are assumptions about supposedly universal and shared values that must be reclaimed to (re)build a fairer society. Yet many NGOs reduce Indigenous Peoples to a token sidebar in policy statements and declarations, as a tragic case study, or otherwise render them invisible or marginal in narratives designed to appeal to liberal audiences.

I call this the "white progressive economic nationalist" position. It obscures and silences long histories of struggle for justice within and against the state by Indigenous Peoples (and other racialized peoples). In the Canadian context, critiques of neoliberal globalization advanced by high-profile NGOs like the Council of Canadians, and the official positions of most major Canadian trade unions exemplify this position and are woven throughout their literature and campaigns. There is little reflexivity about the knowledge on which they base their concepts of social justice and their own roles in reproducing colonial power relations. Largely absent is any genuine acknowledgement of the colonial underpinnings of Canadian state and society, the ongoing denial of Indigenous Peoples' rights to self-determination, and the highly racialized construction of Canadian citizenship and state.

Indigenous Peoples' continued assertions of self-determination and decolonization are a rich but usually overlooked source of theory and critique of both capitalist economic systems and the state itself. Now, transnational corporations often act as new colonial forces alongside older forms of state power. From the corporate enclosure and control of nature through bioprospecting and the imposition of intellectual property regimes, resistance to oil and gas projects, Indigenous and other colonized peoples are at the forefront of both analysis and action against neoliberal capitalism, emphasizing how neoliberal theory and practice commodify all things, are fundamentally predicated on exploitation of people and nature, and embody a colonial mindset. Yet such knowledge and legacies of struggle remain marginalized within global justice networks.

I saw the "white progressive economic nationalist" frame at work rather vividly during the 1997 People's Summit (NGO Forum) on APEC in Vancouver. There, speaker after speaker from Canadian NGOs attacked corporations and the U.S. administration, and identified them as the driving forces behind globalization, yet ignored struggles like that of the Lubicon Cree Nation (in neighboring Alberta) against oil, gas, and timber transnational corporations invading their unceded territory with the complicity of federal and provincial governments. Self-determination for East Timor and Tibet were central agenda items, but no such space was afforded to Indigenous Peoples' struggles within territories claimed by Ottawa. When some of us raised these concerns, session chairs and moderators actively discouraged discussion. Militarism, human rights violations, and undemocratic governments could be challenged if they were Burma, China, or Indonesia. But the fact that a "liberal democratic" government of Canada, like the one that through hosting APEC that year hoped to influence Asian trading partners with "Canadian values" (and indeed many Canadian NGOs agreed) had mounted major armed operations against Indigenous Peoples in the 1990 standoff at Kanehsatake (near Montreal) and again in 1995 at Gustafsen Lake in interior British Columbia, did not warrant mention from the podium. This is hardly surprising—the People's Summit received funding and other forms of support from provincial and federal sources, and many of the Canadian NGOs present enjoyed similar relationships with government. Such silence illustrates the problematic and selective way in which "social justice" continues to be framed by powerful NGO actors.

In January 2005, the Council of Canadians published "The Canada We Want: A Citizens' Alternative to Deep Integration," which repeated the same national narrative of lost innocence and fundamentally egalitarian Canadian society that it had put forward in its APEC activism in the 1990s:

> In order to survive as a country on the northern half of the continent, our ancestors created a narrative of "sharing for survival," which is fundamentally different from the American narrative of "survival of the fittest." Generations of Canadians have been linked together across this huge land with "ribbons of interdependence," such as our national social programs, medicare, our marketing boards, our policies of multiculturalism and bilingualism, and the CBC [Canadian Broadcasting Corporation]. (Barlow, 2005)

Notwithstanding the useful education and mobilization work that such organizations sometimes do, an inability or unwillingness to challenge romanticized and triumphalist nation-building narratives that ignore Canada's founding history of colonization, dispossession, and genocide of Indigenous Peoples is a fundamental flaw inherent in the claims of such organizations to envision and strategize toward a more just society. This praxis reinforces dominant national narratives and what Maori lawyer Moana Jackson (2004) calls the "justificatory mythmaking" (p. 98) apparatus of such supposedly liberal democratic nation-states in relation to both past and present. Instead many of these "alternative" actors claim a positional superiority for the knowledge claims of a small, predominantly white, professionalized NGO/activist elite, which in turn enacts and reinforces what Hage (2003) dubs "white colonial paranoia" (p. 5). Hage, writing of the Australian context, another colonial-settler society, describes this as a constant, often unspoken fear of loss of Europeanness or whiteness through decolonization and Indigenous sovereignty, alongside a rather racialized vision of Canada that, notwithstanding gestures toward "multiculturalism" and "diversity," centers white English-speaking Canadians as the true defenders and agents of the narrative and the "Canadian values" in which many prominent Canadian global justice activists ground their vision of change.

Despite challenges within networks opposed to capitalist globalization in Canada and exposure to explicitly anticolonial frames of antiglobalization activism within Canada and internationally, it seems that NGOs like the Council of Canadians remain resistant to attempts to meaningfully support analyses and action that connect Indigenous Peoples' ongoing struggles for self-determination with opposition to capitalist globalization. The national mythmaking apparatus of purportedly liberal democratic governments such as Canada, Australia, and Aotearoa/New Zealand is of a different and perhaps more difficult order to expose and confront than third-world countries whose governments are more readily seen as undemocratic and unjust in international arenas.

Today, it continues to be the case that acknowledgement and a commitment to confront colonialism and use a lens that sees neoliberal globalization as another wave of colonization are still more likely to be articulated by Indigenous Peoples, landless, small, or peasant farmers' movements, or communities of color directly confronting corporate and state power themselves and allies in smaller activist groups and networks than in larger NGOs. As nation-states restructure under pressure from global capital and the spread of neoliberalism, or enact repressive legislation in the name of "national security" and the "war on terror," so too is their legitimacy questioned by activists advocating self-determination and sovereignty for Indigenous Peoples as a bottom line for a different world order to the one under market capitalism. The sense of betrayal, loss of sovereignty, and despair felt among many nonindigenous people in countries such as Aotearoa/New Zealand, Canada, and Australia affected by free market capitalism also provides windows of opportunity to build solidarity with Indigenous Peoples resisting ongoing colonization. But that work is hard and requires long-term processes of grassroots education and organizing and may not have immediate results.

Knowledge Production in Struggle: Learning from Below

Conceptual resources and analytical frameworks from older histories of resistance against colonialism have valuable lessons for current movements, both in terms of analysis of contemporary global capitalist relations (and their local manifestations) and the dilemmas, visions, and pitfalls of previous struggles. On the other hand, the tendency of many nonindigenous activists and organizations to only support Indigenous Peoples' struggles during visible crises also poses a challenge to building broader anticolonial critiques and long-term alliances and strategies, either locally or internationally. The significance of incremental, long-term political education and organizing work is sometimes difficult to discern, document, or articulate, and yet it is arguably here where lies the hope of challenging dominant practices and discourse, including those that are reproduced in some of the best-known "alternative" and "left" milieus.

An anticolonial lens, drawing attention to past and present features of colonization and decolonization struggles is an important conceptual resource for explicating and historicizing capitalist globalization. D'Souza (2006) argues that it is vital to retrieve, reappropriate, and redevelop the idea of self-determination as a conceptual tool of social transformation in an era in which it has been largely overwritten or excluded from many contemporary analyses. The ideas of those activists that draw explicit links between the advance of global capitalism and older forms of colonialism are often excluded, silenced, marginalized, or otherwise filtered to fit hegemonic positions dominated by NGOs and Northern activist networks. As illustrated above, professionalized NGO advocacy positions often clash with those of social movements and communities for whom these are life-and-death issues (Choudry, 2009; Desmarais, 2007; L. T. Smith, 1999). These movements and communities often challenge the commodification of life (Choudry, 2009, 2010; Kelsey, 1999; McNally, 2002), contesting dominant NGO framings that reproduce capitalist logics. Many NGO documents and platforms reflect a colonial power dynamic—whether it is by positioning the organization as an advocate for peasant farmers' interests in the third world (without any mandate from peasant farmer movements) (Oxfam, 2002; Buckman, 2004), or by perpetuating myths about democracy and social justice in countries like Canada that continue to dispossess and colonize Indigenous Peoples' lands and lives.

As recent mobilizations in Copenhagen around the December 2009 UN summit on climate change, and in Toronto against the June 2010 G-20 meetings illustrate, and as billions grapple with the implications of severe economic and ecological crises, the world of "alternatives" remains a contested arena, ranging from reformist organizations deeply integrated into capitalist relations at national and international levels, to those—mainly in mass-based people's movements—who seek a deep transformation of the capitalist system and its power relations. Indeed, in some cases there is outright hostility and suspicion toward NGOs from mass movements, especially toward those that receive government and/or foreign funding. NGOs have struggled for legitimacy in the eyes of elite institutions and challenges from grassroots organizers and social movements from below, but remain key actors in the national and international political arena. Desmarais (2007) notes the tensions over framing issues, power, strategy, and representation between NGOs and peasant farmers from La Via Campesina in international "civil society" meetings on agriculture. She documents the sometimes dismissive and condescending attitude of some NGOs for whom it is "extremely difficult (if not ideologically impossible) to give up the space they had long dominated" (p. 133).

So, what to do, beyond critique? The richness of activist/movement knowledge and theory, which often occurs in informal spaces and places, is poorly reflected in most of the publications and campaign literature of well-resourced NGOs campaigning on "global justice" or in academic accounts. In many of the "alternative" meetings in which I have participated, debates that questioned hegemonic NGO practices, power, knowledge, mandate, and representation were often shut down or avoided. Yet it was often those moments, when critical voices were raised from the floor of a meeting, that broke through the silences and erasures of supposedly "alternative" NGO meetings. It is partly through recovering, documenting, and validating the knowledge arising from "voices from the floor"—as well as those who are excluded from or refuse to participate in such settings—that we can challenge professionalized NGO forms of knowledge and hegemonic positions within NGO/movement milieus. In doing so, we can contribute to building a body of knowledge and resources for struggle.

In addition, knowledge and histories that emerge from across the colonial divide—from third-world movements and Indigenous Peoples' struggles—offer crucial resources for today's resistance movements. Yet the privileging of Western, professionalized epistemologies of knowledge continues to manifest itself within NGO and activist networks with the reification of "experts" and the dominance of professionalized forms of knowledge such as technical policy analysis. This positions certain kinds of knowledge, individuals, and organizations as authoritative, and devalues or ignores others. This gatekeeping practice and the replication of dominant hierarchies of knowledge is also challenged from the reassertion of grounded grassroots perspectives, non-Western epistemologies and pedagogies of struggle, sometimes within networks dominated by NGOs, and sometimes in entirely separate forums and arenas of struggle.

We need to map the political economy and social organization of NGOs, their ideologies, and claims to representation in relation to donor organizations and broader implications in capitalist relations, asking which perspectives are amplified or suppressed in this process. Such analysis should also examine specific ways in which material support can orient organizations to prioritize institutional survival and maintenance at the expense of mobilization, and account for other ways in which NGO/movement actions may be shaped by material incentives. This has implications for the professionalization of social change and the spread of forms of marketization and competition among NGOs and social movements for mobilization against neoliberalism.

Conclusion

Struggles over knowledge and power are intrinsic to movements for social change. Alternatives arise from struggle, active engagement, reflection, and action. Ultimately, possibilities to think beyond the kinds of ideological pragmatism and constrained dissent practiced by many NGOs need to be anchored in everyday worlds of activist struggle against capitalism and colonialism, and knowledge learned and produced from the ground up. As Rucht (2001) notes, the "shift from radical challenger groups to pragmatically oriented pressure organizations" can lead to a "re-radicalization at the fringes" (p. 220). While changed structures and self-interest in organizational survival may often lead to changed, deradicalized ideologies, this process of institutionalization can drive others to seek different models for their movements.

Note

1. Founded in 1989, APEC brings together twenty-one Pacific Rim governments in a nonbinding, voluntary forum centered around a trade and investment liberalization agenda.

References

Armstrong, E., & V. Prashad. (2005). Exiles from a future land: Moving beyond coalitional politics. *Antipode 37*(1), 181–185.
Barlow, M. (2005, January). The Canada we want: A citizens' alternative to deep integration. Ottawa: Council of Canadians.

Biel, R. (2000). *The new imperialism: Crisis and contradictions in North/South relations*. London and New York: Zed Books.

Bob, C. (2005). *The marketing of rebellion: Insurgents, media and international activism*. New York: Cambridge University Press.

Buckman, G. (2004). *Globalization: Tame it or scrap it?*. Black Point, N.S.: Fernwood.

Burgmann, V. & A. Ure. (2004). Resistance to neoliberalism in Australia and Oceania. In *Globalizing resistance: The state of struggle*, ed. F. Polet, CETRI, pp. 52–67. London: Pluto.

Choudry, A. (2009). Challenging colonial amnesia in social justice activism. In *Education, decolonization and development: Perspectives from Asia, Africa and the Americas*, ed. D. Kapoor (pp. 95–110). Rotterdam: Sense.

———. (2010). Against the flow: Maori knowledge and self-determination struggles confront neoliberal globalization in Aotearoa/New Zealand. In *Indigenous knowledge and learning in Asia and Africa: Perspectives on development, education, and culture*, ed. D. Kapoor & E. Shizha, pp. 47–62. New York: Palgrave Macmillan.

Desmarais, A. A. (2007). *La Via Campesina: Globalization and the power of peasants*. Halifax: Fernwood.

D'Souza, R. (2006). *Interstate disputes over Krishna waters: Law, science and imperialism*. New Delhi: Orient Longman India.

Dür, A., & D. De Bièvre. (2007). Inclusion without influence? NGOs in European trade policy. *Journal of Public Policy 27*(1), 79–101.

Foley, G. (1999). *Learning in social action: A contribution to understanding informal education*. London and New York: Zed Books.

Hage, G. (2003). *Against paranoid nationalism: Searching for hope in a shrinking society*. London and Sydney: Pluto.

Jackson, M. (2004). Colonization as myth-making: A case study in Aotearoa. In *A will to survive: Indigenous essays on the politics of culture, language and identity*, ed. S. Greymorning. New York: McGraw-Hill.

Kamat, S. (2004). The privatization of public interest: Theorizing NGO discourse in a neoliberal era. *Review of International Political Economy 11*(1), 155–176.

Kelsey, J. (1999). *Reclaiming the future: New Zealand and the global economy*. Wellington: Bridget Williams Books.

Khor, M. (2001). *Rethinking IPRs and the TRIPs agreement*. Penang: Third World Network.

Kinsman, G. (1997). Managing AIDS organizing: "Consultation," "partnership" and "responsibility" as strategies of regulation. In *Organizing dissent: Contemporary social movements in theory and practice*, ed. W. K. Carroll, pp. 213–239. Toronto: Garamond.

McNally, D. (2002). *Another world is possible: Globalization and anti-capitalism*. Winnipeg: Arbeiter Ring.

Oxfam International. (2002). *Rigged rules and double standards*. Oxford: Oxfam.

Petras, J., & H. Veltmeyer. (2001). *Globalization unmasked: Imperialism in the 21st century*. New Delhi: Madhyam Books.

———. (2003). *System in crisis: The dynamics of free market capitalism*. Black Point, NS: Fernwood.

———. (2005). *Social movements and state power: Argentina, Brazil, Bolivia, Ecuador*. London: Pluto Press.

Piven, F. F., & R. A. Cloward. (1977). *Poor people's movements: Why they succeed, how they fail*. New York: Pantheon.

Raghavan, C. (1990). *Recolonization: GATT, the Uruguay Round and the developing world*. London: Zed Books.

Rodriguez, D. (2007). The political logic of the non-profit industrial complex. In *The revolution will not be funded: Beyond the non-profit industrial complex*, ed. INCITE! Women of Color Against Violence, pp. 21–40. Boston, MA: South End Press.

Rucht, D. (2001). The transnationalization of social movements: Trends, causes, problems. In *Social movements in a globalizing world*, ed. D. della Porta, H. Kriesi, & D. Rucht, pp. 206–222. Houndmills, Basingstoke: Macmillan Press and New York: St. Martin's Press.

Smith, A. (2007). Introduction: The revolution will not be funded. In *The revolution will not be funded: Beyond the non-profit industrial complex*, ed. INCITE! Women of Color Against Violence, pp. 1–18. Boston, MA: South End Press.

Smith, G. W. (2006). Political activist as ethnographer. In *Sociology for changing the world: Social movements/social research*, ed. G. Kinsman, A. K. Thompson, & K. Tilleczek, pp. 44–70. Black Point, NS: Fernwood.

Smith, L. T. (1999). *Decolonising methodologies: Research and Indigenous Peoples*. Dunedin: University of Otago Press and London: Zed Books.

Townsend, J. G., & A. R. Townsend. (2004). Accountability, motivation and practice: NGOs North and South. *Social and Cultural Geography* 5(2), 271–284.

Watkins, K. (1996). Free trade and farm fallacies: From Uruguay round to the world food summit. Presentation at Manila People's Forum-APEC socio-economic cluster forum, Davao, November 19–20, 1996.

Wood, E. M. (1995). *Democracy against capitalism: Renewing historical materialism*. New York: Cambridge University Press.

CHAPTER 3

Knowledge and Power in South Africa: Xenophobia and Survival in the Post-Apartheid State

Ashwin Desai and Shannon Walsh

In May 2008, South Africa was racked with the worst xenophobic violence since the end of apartheid. In the space of a few weeks, more than sixty people—overwhelmingly migrants from other African countries—were viciously attacked and killed by bands of vigilantes. Tens of thousands of people were displaced, many seeking protection outside local police stations, community and church halls, and temporary, precarious camps constructed throughout the country.

The camps became the most lasting visible reminder of the violence. Hemmed in by barbed wire, the former township residents braved the winter in the Cape and Gauteng. In many cases their possessions were lost and their shacks burned to the ground or taken over by former neighbors. Initially the state responded with denial, or tried to characterize the events as the work of criminals or a "third force" (*Mail & Guardian Online*, 2008), but these explanations were difficult to sustain as images of death mounted in the newspapers. While the state dithered, denied and hoped the "problem" would go away, individual volunteers and faith-based organizations mounted a significant response of support through clothing and food donations (Amnesty International, 2008).

While the violence simmered, the "knowledge industry" of research institutes and nongovernmental organizations (NGOs) attempted to come to grips with both the root causes of the violence and the local sparks that lit it. A bewildering mix of explanations and bald assertions about the perpetrators' identities appeared. Rarely were those subjected to the violence heard from, even though many of them knew their attackers. When their perspectives were solicited, it was rare to discover how they had organized.

Throughout this chapter we explore how the underlying meaning and contradictions that the refugees posed for the South African state were covered over and controlled by a stranglehold on knowledge production that silenced the voices and firsthand accounts of refugees themselves. The xenophobic violence and official responses to it open an important window into South African society and the way power and knowledge work together. They challenge a transition that, while often written in romantic terms, has seen fifteen years of increased inequality and worsening conditions for the poorest sections of society. Contemporary South Africa has moved from apartheid to a liberation where noncitizens, or excluded bodies, are becoming positioned as pools of cheap labor power in competition with those who have struggled for freedom over a lifetime. A discourse of human rights and citizenship is predominantly used to articulate claims for basic services from the state and the heralded "better life for all" promised by the African National Congress (ANC). The specter of the refugee disrupts this fragile relationship. Controlling knowledge production keeps refugee subjectivity invisible and silent, covering over the gap in meaning for both the state and "citizens" who are still hoping to be served.

This chapter discusses how forms of dominant knowledge production within South African social movements, NGOs, and the state exclude migrant and refugee voices, and often emphasize citizenship, legality, and human rights, further reinforcing a nationalist discourse that positions citizens as the only valid bearers of rights, and even justifying the very kinds of attacks it decries. As Agamben (2000) writes, "[i]n the system of the nation-state, so-called sacred and inalienable human rights are revealed to be without any protection precisely when it is no longer possible to conceive of them as rights of the citizens of a state" (p. 3). If you are not a citizen, you are not human.

In this chapter we hear the stories of Abdul and Madondo,[1] two migrants who have been active in working to counter xenophobia, and who have personal experiences of the violence on which they ground their analysis. Abdul uses local knowledge gained through his experiences as a child guerrilla in Somalia to survive in South Africa, while Madondo has had to develop an acute understanding of the local in order to survive in Albert Park, Durban. We also reflect on the research reports that often further excluded non–South

Africans, reinforcing a discourse of nationalism and citizenship that creates ongoing precariousness for non-status people, migrants, and foreigners.

Reflecting on the experiences of people on the ground disrupts the static image of the power of a sovereign state, and broadens and deepens the picture of how knowledge moves and transforms in situations of crisis. Abdul's story, for example, complicates how xenophobia operates in South Africa.

Abdul: Newtown, Johannesburg

Abdul is a twenty-two-year-old refugee living in Newtown, in downtown Johannesburg, who was displaced by the May 2008 violence. Born in Kismayo, Somalia, he was a child soldier who lived through a great deal before he made the dangerous journey to South Africa. He describes his childhood as full of "horror and sadness. I lost my dad, and I always used to see dead people in the streets." He describes himself as "honest and caring. I have good communication skills and come up with influential strategies and I'm a positive thinker. But I was very cruel and heartless when I was young soldier because they used to give us drugs that made us not afraid from any one." Abdul did not have much of a childhood.

> Me, I'm a guerrilla who's been carrying an AK[47] since I was fourteen. Look at this scar on my leg. Once we (me and two friends) were playing soccer. It was 1996 in Somalia. Nearby, two guys were exchanging fire. A bullet hit me through both knees. I didn't feel it then. But I felt it after. I lost a small brother who was playing on the beach. We are like Durban—on the Indian Ocean. He was the youngest. He was playing with a land mine next to the wreckage of a fishing ship. We just heard a sound. When we found him his body was separately apart. He was seven years old.

Abdul fled the violence in Somalia, arriving in South Africa hoping to study and have a better life. He runs a shop amid the flats in Newtown with his uncle, selling goods cautiously through a grated window. In Somalia, he convinced a man from a local mosque to sponsor him so he could get a diploma in sales and marketing. The man agreed, but Abdul was only able to go so far:

> I wanted a better life and to get an education. That's the thing that matters most.... [but] in Somalia things are extremely bad. It's war! One time a bomb dropped right onto our house. My father and my sister were killed in that explosion. The rest of us only survived because we were in

different places. We lost everything. That's why I've such problems with my personal documents.

In South Africa, he dreamed that he could continue his studies, but the reality was quite different:

> My only real job now is to save my life—to stay alive. Then there's the issues of permits and visas and documents. Whether your papers are valid or not, South African authorities will still make it difficult for you to get anything done.

Ironically, the idea of "rights" allows citizens protection while denying rights to noncitizens. Abdul explains how this permits violence and theft to occur with impunity:

> South African people don't think you have rights. Me, I've been robbed by people I know! And you never report it. You will die. All you think about is how to make it not happen again. That's all you can do. There's this one guy—Tsepho.... Once I sold him a can of Lucky Star fish—you know the one to make food for the kids. He gave me R20 [South African rands] and I gave him R6 change. He said, "No, the change is supposed to be R100." I refused. Next thing I know there's an argument. Next, there's a nine-millimeter [pistol] coming out. In the end, they took R5,000 worth of airtime and R2,400 in cash. And, to this day, he still comes to my shop. He smiles at me. He greets me. I greet him. He laughs. He knows there's nothing I can do. I'm Somali. I have no rights. And to add to this scenario, the owner of the shop that we are renting, Adam, his son is a policeman! He knows what's going on and he does nothing about it.

Abdul survives through a variety of tactics. He runs the shop with his uncle, but says he is also a hustler: "I know how to operate. I always change my routes. I always change my times of movement. But still I make it. I know the area." Even under so much duress, Abdul is committed to helping other refugees: "I'm not just committed to a job, but to a community service job—I take people to refugee services when people need me. I go to police stations and court with them."

Abdul explains the current situation for Somali refugees in South Africa:

> At least eighty-seven people have died—Somalis—since May 2008. The worst scenario is the family that was killed in East London just now.

The mom was stabbed 118 times. The daughter of fourteen got repeatedly raped. There's no dad in the family. He's in the UK. The other two kids also got killed—an eight-year-old and a sixteen-year-old.... What's worse is that when these things happen, they torture them first before they kill them. They burn them alive. They hurt them horribly. Like the girl they raped, they also cut her private parts when they were done. In Motherwell, Port Elizabeth, there are more than 6,000 shops owned by Somalis. Last week we just finished the census. There are about 8,000 Somalis living there. In Cape Town it is 12,000.

In May 2008, about 800 Somalis in Acasia, Pretoria, were displaced by the violence. Abdul found himself in one of the temporary camps:

[The Acasia camp] was in a bush, man. Just a bush. I was there two months. June and July 2008. There were eight big tents. As men, first priority is to women and children. We men made shacks out of boxes and blankets. And it was winter. No hot water. Temporary toilets that they don't even clean. We had to go into the forest to "toilet." It was subhuman. Nobody human can live in that situation.... It was zero humanity in those camps.... That's the worst situation I've been in. Those are no living conditions—for anyone.

Much is written about the role of volunteers in the aftermath of the violence, yet there has been almost no mention of the fact that Somalis built their own organization—the Somalia Association of South Africa (SASA)—mainly run by volunteers, of which Abdul is a member. SASA has built a coalition against xenophobia. Abdul is one of the volunteers who facilitates anti-xenophobia workshops in communities, sponsored by the Nelson Mandela Foundation.[2] He collaborates with NGOs and organizations like SASA, finding ways to solidify networks with other refugees, to be safe, to access information, and to contribute to his community. As Abdul reports, "Somalis also always stand together. You can walk into any of their shops and ask for help, and any amount of it, and they will assist you." While this is somewhat ironic coming from a country torn by violence, the circumstances in South Africa are creating conditions for Somalis to look beyond clan and regional rivalries to build networks of support that not only crisscross South Africa, but also move through Somalia into Europe and the United States.

Survival counts on community and the bonds formed between others in the same situation. Abdul talks about his fears, but feels he has no choice but to stay on. He lives in fear and terror, afraid of being killed. Based on his experiences there is no recourse, no protection from violence for people like him. The state and the police are no help. According to Abdul:

> The worst problem we have is the police. I am a member of the Community Policing Forum in three areas: Newtown, Brixton, and Johannesburg. After one meeting I got a lift back to Mayfair. On the way, I saw a Somali guy being chased at night. He was carrying R15,000 in cash. The robbers knew he was going to buy stock for his shop for the next day. And there's nowhere to go for help. Even if a Somali calls an ambulance for an emergency they will never come. The moment they hear the address, you can forget about it. They (all the police and other officials and authorities) know that everybody there (8th Avenue in Newtown) is Somali and you're just a f—in' refugee and you don't deserve to be served. Same with the justice system—there's no pressure to work on the matter, pursue the criminals or facilitate court processes. Nothing. The attitude is: Don't worry. No pressure. It's not important. It's only f—in' *kwerekweres* [derogatory term for foreigner]. The police are so corrupt. They watch incidents happen and don't do anything about it. In fact, they are part of it. They also assault and rob people. You've got to pay money to them all the time.

Abdul bravely continues the fight to survive, deepening the networks he has with other Somalis and finding cracks in the system. Unfortunately, his story is not uncommon, as Eugene Madondo in Durban shows us.

Eugene Madondo: Albert Park, Durban

Some see the May 2008 violence as an episode that is now over. Yet every day on the streets of South Africa xenophobic violence continues. In January 2009, a horrific attack took place in Durban, recounted by Eugene Madondo, who experienced it firsthand.

Madondo grew up in Zimbabwe. When inflation started to get really bad, he opened a stall to sell clothes in a flea market in Chipenge, his hometown. But his venture proved impossible:

> The government launched a state operation, which started destroying illegal flea-markets. Mine wasn't illegal. I was paying money to the municipality for using that space.... Most people were selling food, especially cooking oil, rice, and mealie meal because those things you couldn't find in the shops. They even burnt them down—our flea markets. Then later in 2007, I started to cross the border to South Africa. Why? Because

everything was gone. I was looking for another way to make a living. My market was burnt. I had nothing. No choice.

After leaving Zimbabwe in April 2007, Madondo began job-hunting in Durban. Through a chance meeting, he secured a job as a plumber in Pinetown. In the Albert Park area of Durban there were two antimigrant mobilizations through December 2008 and January 2009, while Madondo was on holiday from work. In one, allegedly instigated by ANC ward councilor Vusi Khoza, locals entered a block of flats in St. George's Street called Jamba House, occupied largely by migrants. They proceeded to throw all of the foreigners out. Jamba House had been the scene of several police raids. Residents allege that police would simply confiscate whatever they had. Lawyers for Human Rights (LHR), an independent human rights NGO subcontracted by the UNCHR (the United Nations refugee agency), had taken up a number of cases after residents produced receipts for the goods. Once the residents were out on the streets, police harassed them until they finally disappeared into the night.

Madondo, aware of the situation, felt uneasy, but thought it was simply the excesses of the festive season. As dusk fell, he took refuge in his room at Venture Africa, another nearby block of flats mainly housing migrants.

> Then came this attack on 4th January 2009. It was on a Sunday at 10:30 pm. I saw a group of people carrying the weapons like bush-knives and knobkerries [clubs]. They were carrying hammers, too. They were blowing the *vuvuzela* [a horn]. They were chanting the slogans in Zulu languages, and some of them were singing. I saw them coming in the direction of the flat where I stay, Venture Africa.

To get to Venture Africa, the mob had to march past the police station. Closed-circuit television footage caught the marchers on the streets. An employee at the LHR office close to Venture Africa repeated that people "heard the mob, saw the mob." The owner of Venture Africa was contacted and phoned the police to intervene. The police told him it was the legitimate work of the Community Policing Forum (CPF).

> [The mob] forced their way up. Six floors.... When it started, I was sleeping, but I was SO scared. I heard the noise of the people screaming. The doors were being broken. They were hitting the doors with the hammers. When I noticed they were even breaking the doors, I thought of running away, so I opened the door, only to see that my neighbor was being thrown from the sixth floor—DOWN! Before I could lock the

door of my room, I was approached by this group of people—a lot of them; there were a LOT! The men were in front, and the women were behind shouting, "*Shaya! Shaya!* [Beat them! Beat them!]" They asked me where I'm from. I told them I was from Zimbabwe. They shouted, "*Shaya kwere-kwere! Shaya!* [Beat the foreigners!]"[3] One of them hit me on the head with a knobkerrie. The blood started to flow. I felt weak. I screamed for help. My attacker was joined by other guys. They hit me with blows on my stomach. They even hit my genitals. I felt so weak. I fell down. I thought it was over. I heard one of them talking in a Zulu language. They said, "Let's throw this dog outside the window," in Zulu. Five of them, they lift me up, pushing me though the window. I tried to hold the window frame, calling for help. They broke the window frames, pushing me down from the fifth floor. Lucky enough, I fell on top of these two other guys who were already dead on the floor. When I was down there, and they saw that I was not dead, they carried on throwing stuff and hitting me with empty beer bottles and all those objects. I didn't even notice these guys were dead. I told them, "Come on guys. There's a storeroom nearby that we can crawl to." The blood was coming too much from my head. I could feel the pain in my spinal cord and my head. I lost my conscious[ness].

Others were pushed too. A group of African immigrants seeking shelter at Albert Park, constantly harassed by police, begged the city to turn it into an official camp. Local NGOs, some in the pay of the UNCHR, tried to convince the group to disperse. When their large tent was removed, they built makeshift shelters out of bin packets.

Meanwhile Madondo made his way back to the scene of death. His room had been ransacked. Many of his fellow tenants had vanished fearing another attack. Madondo sought help from LHR and the Refugee Social Services. The first thing he was asked to do was to go to an identification parade.

I went there, alone. Most of the witnesses who are not victims are scared. They have fear. They think they will be killed by the perpetrators of the attacks, because most of them are out on bail.

On 19th May was the Investigation Parade where I managed to point to some of the perpetrators, including Vusi Khoza, himself. Vusi, what can I say about Vusi? He is so evil. He is lacking a sense of living together. Before the attack he used to come to Venture Africa pointing the fingers [at] the foreigners who live in the flats.... He is not good, that man. And, he doesn't want to repent, or apologize. He is trying by all means to destroy the evidence by killing the state witnesses.

When I pointed him out in the ID parade, he could not see me because of the screen. He was in position fifteen. They said, "Number fifteen, please step forward." When he did that, even though he could not see me, he shook his finger to let me know to feel threatened and have bad feelings. But I don't hate him. All I don't like is how he treats other people, especially the foreigners. All I need is for him, especially him, Vusi, to be trialed for what he did and he must get a sentence.... The situation at home is causing us to leave our home countries. So, if you take refuge in another country and you find out that the situation in that country is even worse than back home, and local people are out to kill you, then where actually do we go?

Madondo survives by relying on the systems of support that he has developed through family and community ties, though through the violence his wife and young son have fled back to Zimbabwe. Vusi Khoza is still on the prowl in the Albert Park area. He remains ANC ward councilor and chairs the Metro Police Civilian Oversight Committee. He claims credit for the Metro Police moving headquarters to what was once the Albert Park bowling club. Pressure is likely to increase on African migrants in the area.

Research Reports and Social Movements: Silencing Migrants

The stories of people like Abdul and Madondo paint a stark picture of the reality for migrants and refugees in South Africa. In the aftermath of the 2008 violence, a number of research reports were published with the aim of providing a deeper understanding of xenophobia in South Africa and to make recommendations to the government to prevent future attacks. In many of the reports we reviewed, the voices of those with relatively high levels of power took precedence over those like Madondo and Abdul who had already lost so much (Graham, Perold, & Shumba, 2009; HSRC, 2008; SAHRC, 2008; SAMP, 2008).

For example, in the one major paper on volunteer work in the aftermath of the violence, no mention was made of the work of African migrants. Instead the paper waxed lyrical on the volunteerism of South Africans. African migrants were erased as active agents, presented as a people without history, passive recipients of the generosity of "private" South African citizens. The paper excludes African migrants from civil society, erasing their self-organizing and celebrating, rather than problematizing the power relationships created between those confined in the camps and those allowed the luxury to be what the paper calls active citizens (Graham et al., 2009).

In another report by the Human Sciences Research Council (HSRC), South Africa's statutory research agency, immediately following the May 2008 violence, researchers made the shocking decision to exclude the voices of African foreigners altogether. They explained that "since the study was focused on gaining an understanding of the views of community members about 'foreigners' and xenophobia, the focus groups recruited only South African citizens" (HSRC, 2008, pp. 4–5). We would argue that these exclusions are themselves another form of violence. In the end, migrants and refugees were hit from various fronts: from the community who rose up to attack them, to state responses that left them in precarious situations in badly serviced camps, and the NGOs and agencies supposedly working on their behalf recommending further exclusions. While NGOs and agency reports are allowed a loud and substantial voice in representing the issues at hand, migrants are once again placed outside the zone of inclusion, confined to bare life,[4] a problem to be dealt with and not warranting consultation or dialogue.

Many assertions made in the HSRC report reinforce and reaffirm oppressive practices toward migrants and refugees. The recommendations are harsh. For example, it calls

> on the government to conduct a national audit on the occupation of RDP [Reconstruction and Development Programme] housing and to take steps to ensure that only South Africans occupy this form of temporary shelter. Non-South Africans are welcome to acquire property through the usual commercial means or to take temporary accommodation that should be provided in designated areas until such time they move into private residence. (HSRC, 2008, pp. 9–10)

The RDP was part of government efforts to redress inequalities bequeathed by apartheid. Through this program, government subsidies are available for the construction of cheap houses, and while more than a million houses have been built, this has barely begun to address the need throughout the country. Does the HSRC report mean that those non–South Africans occupying RDP houses will be evicted? This will only further stigmatize and isolate the poorest African migrants and embolden locals to "take the law into their own hands." It also displays a dangerous lack of understanding about the economic situation of the majority of migrants. The chances of acquiring "property through the usual commercial means" are very slim in a country with more than a million people waiting for housing. The consequence of the HSRC (2008) recommendations would have African migrants living in Bantustans[5] as a permanent feature of the urban landscape, not even allowed to occupy what the HSRC deems "temporary shelter." A new apartheid has emerged.

The report goes on to assert that

> [f]ocus groups describe a "state of siege" environment in regard to survival, access to resources, the state of the economy and to competition from external forces. In this context, the "foreigner" is the nearest "other" against which this sentiment can be expressed. (HSRC, 2008, p. 8)

This finding leads them to suggest:

> The hiring of illegal non–South African citizens in some key sectors of the economy, such as in domestic work and in the construction sector, needs to be terminated. (HRSC, 2008, p. 12)

They recommend imposing fines and other penalties on employers. Attempting to regulate or terminate the gray and informal economies that provide meager means of self-sufficiency will force migrants and refugees further underground where they will become ever more vulnerable. Not only should non–South Africans be driven out of RDP housing settlements and denied employment in key sectors of the economy, the report recommends, but there should also be tighter border controls, heavier policing by the state, and increased state presence in communities. These recommendations would have dire results. Abdul understands this system far too well. As he explains,

> The real enemy of foreigners is the government, not the people. I believe that. A policeman will treat you like a shit.... Government is not promoting dialogue. That camp made out of tents and those green things—portable toilets—even they got destroyed in April this year. They cut the electricity and water in November last year.

The Forced Migration Studies Programme, a research institute at the University of the Witswatersrand, conducted one of the few reports that includes the voices of African migrants (Misago, Landau, & Monson, 2009). It ends with this recommendation:

> As much of the violence is rooted in exclusive local politics, Department of Provincial and Local Government (DPLG) and others should identify and promote positive leadership models and leaders committed to tolerance and the rule of law. In all cases, interventions must be wary

of empowering "unscreened" community leadership structures such as street committees and other forums. (p. 52)

All civic action, the authors recommend, should be legitimized and institutionalized by the state, which should be given the power to decide on who should represent the community's interest. Yet the very same local government structures have been implicated in the violence and have become the target of local communities taking up issues of service delivery.

Migrants and refugees survived through everyday resistance and collective bonds. Recommendations from the HSRC and other such reports will make survival that much more difficult. Abdul and Madondo's stories illustrate how the police and the state are complicit with the xenophobic violence. To ask that same state apparatus to control and repress further as a way to mitigate the impact of xenophobia seems patently misguided.

Part of the problem lies in the fact that the reports seek to make policy recommendations. The authors appeal only to the state, seeing power as solely existing in formal, institutional mechanisms. Consequently, people in the communities are written out of the story. When their voices *are* heard, it is as sources of "intelligence" or knowledge for the state. Funders also demand recommendations that are directed to the state, with NGOs acting as partners or midwives, rather than directing power and resources to more organic community-based movements.

Discussion

The recent increase of economic migrants and refugees in South Africa has a strong parallel to the world that Harold Wolpe described in his essay *Capitalism and Cheap Labour Power in South Africa* (1972). Wolpe described the apartheid system as not only one of racial exclusion, but also one of domination and control intrinsically bent on reproducing and elaborating capitalist relations by maximizing cheap labor with little financial burden on the state. In contemporary South Africa, it is a cheap labor pool *without* rights *within* a cheap labor pool of black South Africans, once again creating a situation in which some members of society are denied rights, without recourse to the state, subjected to bare life, and in this case policed through violence *both* by the state *and* by other poor South Africans who see themselves as bearers (and possible beneficiaries) of certain rights and concessions.

The *deserving citizen* discourse that shapes contemporary social movements in South Africa has tended to obscure the nature of the political, since many post-apartheid mobilizations have focused on service delivery as the main object of action, and mix tactics between courts, councilors, and the streets with a human rights discourse (Robins, 2008; Sinwell, 2009). Ultimately this has often been translated into requests for concessions from the state rather than challenges to the foundations of capitalism and neoliberalism at work in South Africa. Articulations toward nationalism and citizenship, as Hart (2007) notes, are not cynical projections emanating from a state attempting to reinforce its power, but are deeply rooted in the liberation movement, invoking a collective struggle history both from the current Zuma-led ANC government and other state apparatuses, and within social movements and other everyday practices. It is a sad irony that the evocation of a history of racial oppression may be a large aspect of the current nationalist project that is so full of exclusions. The dominant developmentalist approach does not significantly challenge the ANC's neoliberal project and the conditions that cause migrancy, and issues like water, sanitation, and housing. As Luke Sinwell (2009) writes in his analysis of the politics and class consciousness of the Anti-Privatization Forum (APF),[6]

> community-based movements which base their critique of development on claims of corruption are faced with serious limitations in terms of their ability to create an alternative to the ANC's neoliberal or class-based project.... The local critique does not seek to alter the underlying structures of social change that bind people into a state of poverty in the first place. (Sinwell, 2009, p. 5)

As Sinwell (2009) argues, development critiques limited to the local level can "obscure, and indeed sustain, broader macro level inequalities and injustice" (p. 140).

To understand more clearly how this enclave *within* an enclave operates through the circulation of knowledge and power at multiple, simultaneous levels, we must challenge vertical topographies of power (Ferguson, 2006) that position civil society and social movements at the local level, with the state far "above." Such a limited vantage point allows ethnographic richness and the terrains of resistance to be lost. In this case, we must challenge what we mean by "local" versus "state" knowledge. In the xenophobic violence, one group of "local" poor were positioned as policing agents against "local" migrants. As Ferguson (2006) writes,

> The policing of the border is intimately tied to the policing of Main Street in that they are both rituals that enact the encompassment of the

territory of the nation by the state. These acts represent the repressive power of the state as both extensive with the boundaries of the nation and intensively permeating every square inch of that territory. Both types of policing often demarcate the racial and cultural boundaries of belonging and are often inscribed by bodily violence on the same groups of people. (p. 110)

The policing of Main Street within the shack settlement by citizen "locals" in line for government housing is an enactment of legitimation and territorialization of the state. The challenge posed by the commodification of basic services, evictions, and proper housing has created radical refugee subjectivities within South African social movements that have contested the threat to bare life posed by the neoliberal takeover (Robins, 2008; Sinwell, 2009). However, those subjectivities and the movements that have emerged have often used a language of "rights" to assert their basic claims on bare life. In the context of a continent in severe economic distress, with a powerful pull that draws migrants seeking "a better life" into South African territory, and whose claims to oppression and exploitation are no less stark than those of black people who happen to have been born within the country's borders, such repression-oriented, nationalist thinking that excludes voices like Abdul and Madondo will only deepen xenophobic attitudes.

Ultimately the forms of knowledge production that we discuss here in how xenophobia has been represented and dealt with by NGOs and state research reports show the continuing paradox that the refugee as a subject poses for the sovereign state.

While pronouncements of human rights abound,

[w]hat is essential is that each and every time refugees no longer represent individual cases but rather a mass phenomenon (as was the case between the two world wars and is now once again), these organizations as well as the single states—all the solemn evocations of the inalienable rights of human beings notwithstanding—have proved to be absolutely incapable not only of solving the problem but also of facing it in an adequate manner. The whole question, therefore, was handed over to humanitarian organizations and to the police. (Agamben, 1996, p. 160)

Further reinforcing "foreigners" as nonhuman or exceptional through the way knowledge forms and is disseminated underpins the very xenophobia that exercises itself violently on the ground, and is deeply connected to the various institutional players we have discussed here. Revealing everyday forms of organizing, local knowledge, and survival strategies may help problematize this hegemonic knowledge production and reveal the very real inadequacies and violence of official human rights discourse in the age of the refugee.

Notes

1. Interviews were conducted in 2009 with Eugene Madondo, a Zimbabwean migrant, at Albert Park, Durban, South Africa; Ahmed Omar Abdikadir (Abdul), a Somali refugee, in Newtown, Johannesburg, South Africa; and an anonymous legal officer with Lawyers for Human Rights in Durban. Thanks to Marian Murray for help with the interviews.
2. This foundation was set up in the aftermath of Mandela's one-term presidency and was involved in a myriad of activities from research to raising money for a children's fund, with its primary focus on *ubuntu*, or reconciliation.
3. Attacks on African foreigners in the past decade and a half were accompanied by a heightened language of hysteria and demeaning "othering" of African immigrants. The most obvious manifestation of this was the "normalization" of the word *Makwerekwere*. Nyamnjoh (2006) vividly unpacks the significance of this naming:

   > To demonstrate that these "illegals" clearly have little to offer, South African blacks, perhaps reminiscent of the Boers who named the local black communities "hottentots" to denote "stutterers," deny black African migrants an intelligible language. All they claim to hear is "gibberish"—a "barbaric" form of "stuttering"—hence the tendency to classify them as *Makwerekwere*... as used in South Africa it means not only a black person who cannot demonstrate mastery of local South African languages but one who also hails from a country assumed to be economically and culturally backward in relation to South Africa. (p. 39)

4. Bare life (Agamben, 2005) refers to a state of being in which the individual is separated from their social being and particular qualities to be confined to a mere body to be managed.
5. Bantustans, or black African homelands, were created during apartheid to house Africans in "tribal" states and deny them rights in "white" South Africa.
6. The Anti-Privatisation Forum (APF) was formed in 2000 in Johannesburg in the context of the commodification of basic services. It fights various anti-privatization struggles from university campuses to the privatization of electricity and water. Mainly based in Soweto, there are now branches throughout South Africa and affiliates in unions, student groups, and the broader left (see http://apf.org.za).

References

Agamben, G. (1996). Beyond human rights. In *Radical thought in Italy: A potential politics*, ed. P. Virno & M. Hardt, pp. 159–165. Minneapolis: University of Minnesota Press.

———. (2000). *Means without ends: Notes on politics.* [Trans.] V. Binetti & C. Casarino. Theory out of bounds, *20*. Minneapolis: University of Minnesota Press.

———. (2005). *The state of exception.* Chicago: University of Chicago Press.

Amnesty International (2008). *"Talk for us please": Limited options facing individuals displaced by xenophobic violence.* London: Amnesty International.

Ferguson, J. (2006). *Global shadows: Africa in the neoliberal world order.* Durham, NC: Duke University Press.

Graham, L., H. Perold, & R. Shumba. (2009). Volunteering for social transformation? Understanding the volunteer response to the xenophobic attacks of 2008: Implications for democracy in South Africa. *Volunteer and Service Enquiry Southern Africa (VOSESA).* Retrieved from http://www.vosesa.org.za/publications3.asp.

Hart, G. (2007). Changing concepts of articulation: Political stakes in South Africa today. *Review of African Political Economy* (ROAPE) 34:111, 85–101.

Human Sciences Research Council (HSRC) (June 2008). *Citizenship, violence and xenophobia in South Africa: Perceptions from South African communities.* Pretoria: HSRC Democracy and Governance Programme. Retrieved from www.hsrc.ac.za/Document-2807.phtml.

Mail & Guardian Online (May 23, 2008). Xenophobia: A special report. *Mail and Guardian Online.* Retrieved from http://www.mg.co.za/specialreport/xenophobia.

Misago, M., L. Landau, & T. Monson. (February 2009). Towards tolerance, law, and dignity: Addressing violence against foreign nationals in South Africa. Arcadia, Republic of South Africa: International Organisation for Migration (IOM).

Nyamnjoh, F. (2006). *Insiders and outsiders: Citizenship and xenophobia in contemporary Southern Africa.* Dakar: Codesria Books.

Robins, S. L. (2008). *From revolution to rights in South Africa: Social movements, NGOs and popular politics after apartheid.* Suffolk/Pietermaritzburg: James Curry/University of KwaZulu-Natal Press.

Sinwell, L. (2009). The politics and class consciousness of the anti-privatisation forum (APF) in Alexandra: Critiquing class or corruption? Unpublished paper.

South African Human Rights Commission [SAHRC] (2008, July 22). *SAHRC report on refugee camps: Blue Waters, Harmony Park, Silverstroom, Soetwater, and the Youngsfield military base.* Johannesburg, South Africa: South African Human Rights Commission.

South African Migraticn Program [SAMP] (2008). Immigration, xenophobia and human rights in South Africa. *Southern African Migration Policy Series,* no. 22. Queen's University, Canada & Le Cap: SAMP & IDASA (Institute for Democracy in South Africa).

Wolpe, H. (1972). Capitalism and cheap labour power in South Africa: From segregation to apartheid. *Economy and Society 1*(4), 425–456.

CHAPTER 4

On the Question of Expertise: A Critical Reflection on "Civil Society" Processes

Robyn Magalit Rodriguez

In October 2008, the Philippine government hosted the Global Forum on Migration and Development (GFMD) in Manila. Initially convened after a 2006 United Nations (UN) High-Level Dialogue on International Migration and Development, the GFMD, while not formally part of the UN process, is aimed at providing a venue for labor-receiving and labor-sending countries to trade strategies around instituting temporary labor migration programs (TLMPs). Pegged as a "win-win-win" for both sets of governments and migrants themselves, these programs are being celebrated as the best solution to labor-receiving governments' demand for cheap foreign workers to whom they are unwilling to extend full citizenship rights, to labor-sending governments' need to address domestic unemployment, and to bolster foreign exchange reserves, and to migrants' and their families' need for livable wages.

In addition to meetings of government officials, the GFMD instituted a series of civil society meetings to putatively represent the concerns of migrants themselves. Grassroots migrant activists, however, claimed that the GFMD was in fact the "global forum on modern-day slavery" and organized a parallel meeting, the International Assembly for Migrants and Refugees (IAMR), to counter the GFMD government meetings. Perhaps more important, grassroots migrant activists used the IAMR as a venue to also counter the GFMD official civil society meetings. Through the Assembly, activists claimed that migrants would come together to "speak for themselves."

This chapter uses a case study of the struggles around the Manila GFMD to engage a broader critique of the role of officially recognized civil society actors as representatives of ostensibly broader publics in intergovernmental spaces. Here, I am interested in the ways in which particular civil society actors, academics, and NGOs are mobilized as migration "experts" and the political ends to which this "expertise" is used in the GFMD process. I am especially concerned about how nonstate actors are implicated, wittingly and unwittingly, in legitimizing states' neoliberal agendas. At the same time, I explore the sets of knowledge deployed by grassroots migrants and the radically different political ends to which they are mobilized.

The GFMD and "Migration as Development"

At the 2006 UN High-Level Dialogue on International Migration and Development, the UN president declared, "[t]here was widespread support for incorporating international migration issues in national development plans, including poverty reduction strategies" in the dialogue.[1] Indeed, the UN's move to hold the dialogue is part of a resurgent trend among international organizations to promote migration from developing countries as a developmental strategy. The idea is that developing states benefit economically from migrants' remittances as well as from the potential skills and technology transfers that migrants offer upon their return home. Although not formally a UN body and therefore a purely voluntary endeavor, the GFMD nevertheless managed to assemble a wide range of countries from around the world to meet in Belgium in 2007, Manila in 2008, and Athens in 2009.

Temporary labor migration programs appear to be the policy initiative most favored by government participants of the GFMD. TLMPs are believed to best exemplify developmentally oriented migration policies. TLMPs are also said to "optimize benefits for migrant workers, employers, source countries and destination countries" (Government of Bangladesh and Government of Canada, 2008). In short, under TLMPs, everyone supposedly wins.

Ziai and Schwenken's (2009) research has documented that the notion of "migration as development," as embodied by TLMPs as being propagated widely by international organizations beyond simply the GFMD. They note, for instance, that the World Bank alone published ten books and numerous articles on the matter in recent years. Meanwhile, the International Organization for Migration has also jumped on the "migration as development" bandwagon with its introduction of the International Migration and Development Initiative (IMDI) in 2006. Ziai and Schwenken argue that the

TLMP must be understood as a neoliberal strategy that places the responsibility for "development" squarely on the shoulders of migrants themselves. TLMPs at once allow employers to exploit foreign workers, absolve developing states from introducing truly redistributive developmental policies, and relieve labor-importing states from extending the full benefits of citizenship to immigrants. They argue that TLMPs serve

> the interest of firms and service enterprises in industrial countries in a cheap and flexible labour force. As far as this interest is compatible with the interest of certain people in the periphery willing to migrate, the IMDI and the other initiatives surely are bound to have positive consequences for this group. It is clear, however, that the right to migrate will only be granted to those that can be employed or exploited in the market, and presumably only as long as their labour is needed in the countries of destination. (Ziai & Schwenken, 2009, p. 12)

In other words, TLMPs might in fact offer a "win-win-win" situation for employers, labor-receiving countries, and labor-sending countries. Ultimately, however, migrants and their families lose out.

Migrants who work under TLMPs typically do not have a guarantee of stable employment, as theirs are only short-term contracts. Their livelihoods are characterized by a great degree of precariousness, especially when the reason for their overseas employment is because their employment options at home are already quite thin or their earnings are insufficient to support their families. As neoliberalizing developing states cut deeper into state programs and services, including public education and health, families are forced to pay out of their own pockets for basic amenities. Moreover, the temporary nature of their jobs prevents migrants from being able to advance within a firm, to demand higher wages with more seniority, or to accumulate retirement savings. As foreigners, migrants' rights are often severely limited. They are prevented from being able to join trade unions and thereby from fighting for better working conditions. They suffer the daily indignities that often come with occupying the lowest rungs of the occupational ladder as racialized workers in their countries of employment. Meanwhile, migrants' family members suffer from the long absences of their loved ones.

In Manila, the GFMD focused especially on the Philippine TLMP "model" for its putative rights-based approach to migration. According to one policy paper discussed among states at the meeting,

> The Philippine "life-cycle" approach to fostering and supporting the Overseas Filipino Worker programme is a useful model for "protection

beginning at home," which is then reinforced through negotiated, rights-oriented partnerships with both host countries and other non-governmental stakeholders. (Government of Philippines and Government of United Arab Emirates, 2008)

Though a "rights-based" approach to TLMPs may on the surface appear to resolve the problems inherent to such programs, any veneer of "protection" that the Philippine model might offer was the result of hard-fought militant grassroots struggles by migrant workers and not by any foresight on the part of the Philippine state.

Led by an international alliance of migrant workers, Migrante-International, Philippine migrants mobilized transnationally to protest the Philippine state's failure to intervene in what many believed was the wrongful execution of a Filipina domestic worker, Flor Contemplacion, by the Singaporean government in 1995. According to many accounts, these were among the biggest mobilizations by Philippine citizens (both in the Philippines and around the world) since "People Power" brought down the Marcos dictatorship a decade earlier. These mobilizations forced the Philippine state to introduce Republic Act (RA) 8042, the Migrant Workers and Overseas Filipinos Act of 1995. As the GFMD policy paper cited above suggests, the Philippine state has instituted programs that are supposed to offer "protection beginning at home." However, my own research of these programs reveals that they are less about protecting migrants and more about disciplining them to be compliant workers (Rodriguez, 2010).

A superficial reading of RA 8042 might suggest that it does, in fact, offer much in terms of migrants' protection—"at home" and abroad. In actuality the state generally fails to enforce these protections. On the rare occasion that it does enforce them, it does so only after migrants have mobilized to fight for them. Protests against, not "partnerships" with, labor-receiving governments or NGOs, for that matter, are what have led to migrants' protections.

Nevertheless, the formal policy statements of the Philippine state—not necessarily its practices—are being touted as a model to be emulated by other states. The Philippine state attempts to present its program as a rights-based, and therefore modern and progressive one, in its bid to earn a place of legitimacy in the world order. As scholars have long argued, recognition in the pecking order of nation-states is only given over to those states that appear to exhibit some evidence of rationality and governance by the rule of law.

"Civil Society" and the GFMD

The GFMD formally holds a civil society meeting alongside the official meeting to allow NGOs and other nonstate actors to engage in the topics taken up by the government representatives in the GFMD proper. The GFMD Civil Society dialogue in Manila was particularly focused on the theme of "rights." As stated in its summary document,

> We see the challenge to develop global architecture for recognition, respect, rights and protections for migrants as the responsibility of the UN and no less urgent than the need for transparent global governance of the financial system or that required to reduce carbon emissions.[2]

According to the organizers,

> 220 delegates from all over the world, representing concerns for some 200 million migrants met at the second Global Forum on Migration and Development in Manila to consider the rights and protections of migrants, the expansion of legal avenues for migration and the challenge of coherence within nations and across borders.[3]

To become a delegate and fully participate in discussions, groups and individuals were required to submit applications to the organizers. Others who were not accepted as delegates were assigned observer status. Yet several aspects of the delegate selection process raise questions about whether, in fact, migrants' concerns were adequately represented.

To begin with, the very constitution of the Civil Society organizing committee was problematic, as it was constituted predominantly by representatives from businesses that profit from international migration. One of the key forces behind the Civil Society committee was the Ayala Foundation, which had three members on the fifteen-member committee. According to its Web site, the Ayala Foundation, "is a nonstock, nonprofit organization that serves as the sociocultural development arm of the Ayala Group of Companies (AGC)."[4] Among the AGC's most significant business holdings are Ayala Land, Inc., Bank of the Philippine Islands, and Globe Telecom. Perhaps not surprisingly, migrant workers are a key market targeted by Ayala Land, Inc., a real estate development company. The Bank of the Philippine Islands has branches around the world to handle migrants' remittance sending[5], while Globe Telecom sells phone cards for migrants wishing to call relatives in the Philippines.[6] Migration has clearly been a "win" for the Ayala Group of Companies. That its nonprofit wing was such a central actor in the Civil Society coordinating body is suggestive of the sorts of interests ultimately being advanced in the Civil Society meeting. If the Ayala Foundation served the firm's interests by proxy, other businesses profiting from migration had representatives serving on the Civil Society organizing committee without the same sort of pretense. These included one representative from SGV and Co. (a professional services company),[7] two representatives of the Magsaysay Maritime Corporation (a labor recruitment firm),[8] and one representative from the DCL Group of Companies (another major labor recruitment firm).[9] In all, there were seven business representatives on the Civil Society organizing committee.

The remaining three committee members were representatives from different church-based migrant-serving NGOs, a representative from the Alliance of Progressive Labor, a coalition of Philippine trade unions linked with the center-left Akbayan party, and a representative from Migrant Forum Asia, a regional network of NGOs. Finally, there was a representative from the Economic Resource Center for Overseas Filipinos, which describes itself as a nonprofit corporation that facilitates "alternative investment directions for remittances."[10] The broader international advisory committee included more conventionally defined "civil society" actors, including the International Trade Union Confederation (ITUC).

It is not entirely clear who may have comprised the delegate selection committee (although it was the Ayala Foundation that issued the invitations and to which applicants sent their applications). However, those who were permitted to participate in the Civil Society meetings were ultimately drawn from the ranks of NGOs, including migrant-serving organizations and diaspora groups. Representatives from trade unions and academics, including myself, were accepted as delegates. There was even a representative of financial services and communications corporation Western Union who served as a delegate.

Notably, many of the NGOs with delegates at the Civil Society meetings—like the International Foundation of Alternative Financial Institutions, Migrant Rights International, and Global Alliance Against Traffic in Women—are groups with official UN Economic and Social Council (ECOSOC) status. ECOSOC status is conferred by the UN to select NGOs who are then granted the privilege of attending, observing, and in some cases participating in UN processes. These NGOs are a privileged set of actors who can enjoy an audience with government officials to some extent, and who are networked with officials in the UN system in ways that unaffiliated NGOs are not. The Civil Society meetings of the Manila GFMD, therefore, included NGOs who already enjoy considerable access to states.

To what extent NGOs such as those that secure ECOSOC status can adequately represent migrants' issues, however, is debatable. In her research of NGOs focused on women's issues in Latin America, for instance, Sonia Alvarez (1999) finds that they operate as gender "experts" to governments and multilateral organizations like the UN, rather than actually advocating for women's rights. Representatives from these NGOs are believed to represent broader social constituencies, though it is not always the case. Khagram, Riker, and Sikkink (2002) offer a definition of NGOs that attempts to distinguish them from grassroots, social-movement-oriented groups:

> NGOs are private, voluntary, nonprofit groups whose primary aim is to influence some form of social change. Generally, NGOs are more formal and professional than domestic social movements, with legal status and paid personnel. (p. 5)[11]

Indeed, the critical difference between many NGOs and grassroots migrant organizations has to do with leadership, membership, and methods for engaging in social change. NGOs are formally constituted and often officially recognized or registered by the governments where they operate, generally relying on donors (government and private) and staffed by professionals who then craft programs to service specific constituencies. Often funders can play a role in limiting the sorts of activities NGOs can engage in. Grassroots migrant organizations, however, are often led by the constituencies they also serve and are membership-based organizations. They may or may not depend on external funding, but they are often tied to broader social movements engaged in struggles for large-scale structural change.

Based on my observations, I would argue that Alvarez' critique of NGOs could be made of the NGOs permitted to participate in the GFMD Civil Society meetings in Manila. These NGOs and other civil society actors have emerged as migration "experts"; the "expertise" they offer actually appears to advocate states' interests rather than the interests of migrants. Here, Jad's (2007) cautionary approach to NGOs, based on research of women's NGOs in Palestine, is also important to heed: "While 'NGO' may be a synonym for 'progressive' and 'participatory' among the well-meaning supporters of well-known international NGOs, such associations are wishful thinking at best and illusory at worst" (p. 624).

For example, International Trade Union Confederation president Sharan Burrow, who presided over the Civil Society meetings, expressed support for the so-called rights-based TLMPs being discussed by states participating in the GFMD discussions. Indeed, while the panel topics for the Civil Society meeting matched the panel topics for the government meeting ostensibly to generate direct (perhaps even critical) responses by nonstate actors to the governments' policy discussions, the papers put forward by civil society members appeared to make recommendations that were no different from the governments'. For instance, just as governments lauded the Philippine "model" for its "rights-based" approach, civil society participants did the same. One former Philippine official, Patricia Santo Tomas, who was once the government's secretary of the department of labor and employment, served as an "expert" for the Civil Society meeting. Her 2008 paper entitled "Protecting Migrant Workers: A Shared Responsibility" differed little from the paper produced by government officials. Similarly, she highlights the Philippines' supposed regulatory framework for the protection of migrant workers. Though she purportedly aimed to point out the "gaps that must be addressed to ensure the improved welfare and working conditions of migrant workers," (p. 1) Santo Tomas ultimately concluded that the gaps lie less with the fundamental problems with TLMPs, but rather, "are a function of circumstances which are often cultural rather than deliberately criminal" (p. 19). She explains how, for example, "domestic helpers expect to work no longer than 10 hours and expect a day off once a week. These practices are not recognized in some societies" (p. 21). In short, for her, long working hours are a consequence of different cultural understandings of the workday in different receiving countries rather than a violation of migrants' rights. Santo Tomas' former government experience makes her an "expert" on migrant rights, but the same status is not extended to the very people who have experienced terrible working and living conditions: migrants themselves. As I will discuss later, representatives from Philippine migrant organizations critical of the Philippines' so-called rights-based migration regime were excluded from the Civil Society meetings.

Furthermore, among the groups of "experts" setting the terms of discussion in the panels of the "Civil Society" meeting were not only former government officials, like Santo Tomas, but also representatives from other intergovernmental, multilateral institutions like the Organization for Economic Co-operation and Development (OECD), and the International Organization for Migration (IOM), which have generally supported the global propagation of TLMPs. In addition, "experts" were drawn from mainly U.S.- and U.K.-based migration think tanks. Only a small handful of "experts" came from NGOs, and none represented grassroots migrant organizations. In effect, the terms of discussion for the Civil Society meetings were delimited by actors already invested in TLMPs or by academics who, while not necessarily invested in these programs, were not scathingly critical of them either. TLMPs were not subject to any real debate. Moreover, voices of the very people who already struggle to live dignified transnational existences under temporary labor migration regimes—migrants themselves (in those cases when migrants actually served as delegates to the meeting)— were rendered marginal to the voices of experts. Instead, those who led and participated in the official Civil Society discussions were clearly "professional" NGO staffers. During the workshop titled "Voices from the Regions: Regional Perspectives, Essentials and Recommendations in International Migration and Development," the chairperson facilitating the discussion, Ellene Sana, executive director of the Center for Migrant Advocacy, actually admonished workshop participants to limit their discussions to "doable" or "actionable" policy recommendations as opposed to sharing narratives about migrants' experiences, or if they were in fact migrants themselves, from sharing their personal stories of migration. However, the notion of limiting discussion to "actionable" policy is problematic because it fails to recognize that "actionable" policy is precisely the kind that is confined to the imperatives defined by GFMD states. NGO professionals foreclosed the possibility for participants to offer up more radical visions of change.

The most glaring absence among the Philippine representatives was a representative from Migrante-International, a group that perhaps has the most to say about the Philippine "model." It is a transnational alliance of grassroots Philippine migrant organizations with members in nearly every region of the world. Despite its long history of championing migrants' rights and contesting the Philippine government's migration policies, and its critical role in compelling the Philippine state to pass the very law that was held up as the rights-based model of the GFMD, Migrante was denied an opportunity to send a delegate. Its application, according to former chairperson Connie Bragahas Regalado, was rejected. The venue where the Civil Society dialogue was held was heavily guarded by the Philippine National Police and entry into the dialogue was strictly limited to those granted official delegate status.

Interestingly, many of the groups that were admitted had also participated in a preconference entitled "Peoples' Global Action on Migration, Development, and Human Rights." The conference billed itself as a venue for expressing resistance, struggle, and opposition through which to "assert the migrants' and people's perspectives and human rights"[12] against the GFMD. It was convened by the Philippine Working Group on GFMD and Migrants' Rights International. Many of the organizations that convened this conference were also delegates to the Civil Society days and were actually well placed as members of the Civil Society Philippine Organizing and International Advisory Committees, serving as resource persons and even chairing workshops. Arguably, those of the Peoples' Global Action on Migration, Development and Human Rights (PGAMDHR) who were simultaneously positioned in the official Civil Society meetings could have asserted the moral and representational authority of the premeeting to direct discussions in a genuinely critical direction. They did not. Those NGOs and other "experts" who were incorporated into the process seemed to have merely served to legitimize states' imperatives for the GFMD rather than using their access to fully give voice to migrants' concerns or critiquing the official "migration as development" framework. Nandita Sharma's (2005) critique of the antitrafficking discourse as articulated by some women's movements, governments, and the UN is applicable to the GFMD Civil Society process. She argues that, "[t]he ideology of anti-trafficking does not recognize that migrants have been displaced by practices that have resulted in the loss of their land and/or livelihoods through international trade liberalization policies, mega-development projects, the loss of employment in capitalist labor markets or war" (p. 94). Advocating a "rights-based" approach to "migration as development" as was done by the delegates of the GFMD Civil Society process fails to challenge how structures of neoliberal globalization and the inequalities it produces between countries ultimately fuel international migration.

GFMD: Global Forum on Modern-Day Slavery?

Across the street from the official Civil Society venue, at a hotel that was quite modest by comparison with the posh Heritage Hotel that housed the Civil Society delegates, self-organized, grassroots migrant groups gathered in an alternative, counter-meeting at the International Assembly of Migrants and Refugees (IAMR), through which "the genuine voice of the migrants could be heard."[13] Unlike most of those participating in the official meetings, they took a sharply critical stance of the GFMD, characterizing it as the Global Forum on Modern-Day Slavery.

More than a hundred delegates from around the world attended the IAMR. Testimonies featured prominently in the IAMR, as migrants and their family members were encouraged to share their struggles. The crowd of several hundred migrants was transfixed by Elvira Arellano's harrowing account of her struggle to seek sanctuary in a Chicago church to evade agents of the Department of Homeland Security as she spoke in a street demonstration organized by the IAMR. Elvira's audience jeered at her experience of the brutal inhumanity of the U.S. immigration system, which was ultimately successful at wresting her and her U.S.-born son from the church and deporting them to Mexico. At the close of her short speech, Elvira called on audience members to join together in the fight for migrants' rights and roused people to chant, "Sí, se puede [yes, we can]", an immigrant-rights anthem that long predates Barack Obama's most recent invocation. Elvira's and other migrants' stories shared at the demonstration were meant to raise consciousness to build relationships and to galvanize and inspire people for collective action. Knowledge was harnessed here as a tool for social justice activism and mass mobilizations to make demands of governments and to hold them accountable.

The centrality of public testimonials to the IAMR for the purpose of mobilizing migrants in collective action stood in stark contrast to the activities of NGOs and other civil society actors in the GFMD's official Civil Society Days, where testimonials were explicitly excluded from workshop discussions, and where, when a migrant's testimony was called for, it was deployed in what looked like a kind of publicity stunt. For instance, at the conclusion of the "civil society" conference, in the closing plenary session, "The Interface Between Civil Society and Government," when recommendations from all the prior workshops were to be summed up and presented to government officials, ITUC president Sharan Burrow called upon a Filipina migrant woman employed as a nurse in Italy to offer her policy recommendations to officials. The move caused a bit of a stir because the woman was already making her way out of the conference venue. She had to run back to address the officials and delegates and was breathless from running. She clearly did not expect to be speaking and was unsure of what to say. She then proceeded to narrate her experiences of migration and her opinions about what government officials ought to address in terms of migration policy. Her recommendations were purely individual recommendations. They did not reflect the collective discussion of delegates that had take place over the course of the conference. It was later revealed that a group of preselected delegates (the selection process was concluded even before the Civil Society meetings) would address government officials in a private meeting, presumably to offer up the collective discussions of the Civil Society meetings, but away from the scrutiny of the larger body.

Indeed, if the IAMR was a venue through which migrants could share their stories not only to provide real-life testimonies of the violence of TLMPs, but also to cultivate collective identities as migrant workers, it was also a venue through which migrants could cultivate alternative forms of knowledge about the GFMD. As Eni Lestari, an Indonesian migrant domestic worker activist and secretary general of the newly formed International Migrants Alliance (IMA), which helped organize the demonstration, put it, "For many years, many have spoken on our behalf. This time, we will speak for ourselves." For IMA, the street demonstration and the counter-GFMD conference that it organized was for the purpose of expressing migrants' concerns more authentically. These were sites for the production of alternative forms of knowledge about the GFMD to counter not only the knowledge produced by governments, but also that produced by mainstream, professional NGOs who were permitted to participate in the official "civil society" meetings.

Because the IAMR was organized outside of the official Civil Society process, migrants could express analyses that were not constrained by the interests of those defining the Civil Society meetings' framework. Knowledge production at the IAMR, whether occurring on the street or in more formalized conference settings becomes the site for "meaning work" (della Porta, Andretta, Mosca, & Reiter, 2006, p. 62).

The call of the IAMR conveners was not for reforms of TLMPs but for their rescission. For them, development should not be addressed through migration policy, but rather through initiatives that alleviate poverty and create decent employment at home. They argued:

> Though it professes to not substitute migration for genuine development, it is evident in the GFMD agenda and process that what it does aims [sic] for is to utilize migration as a cover up for the destruction that neoliberal globalization has heaved to the people's lives. (Asia Pacific Mission for Migrants et al., 2008)

IAMR participants expressed an uncompromising stance on the neoliberal logic of the GFMD, unlike their counterparts in the Civil Society meetings. Though many of those in the latter meeting were positioned to shape the discourse around the GFMD and had expressed a similar critique of the GFMD during the PGAMDHR conference, they remained silent on the issue when afforded the opportunity to face government representatives. Instead they oriented themselves toward reforming the GFMD through an agenda already sanctioned by states rather than struggling to alter the terms of discussion.

Although the IAMR certainly attempted to give primary attention to migrants themselves, they also included "experts" in their meetings. Their use of "experts," however, served different ends. Jorge Bustamante, a sociology professor who also serves as UN special rapporteur on migrants' rights, addressed the IAMR and expressed his critique, as a migration scholar, of the notion "migration for development." He argued, similar to IAMR conveners, that it is the wrong approach as it makes migrants' responsible for national development while absolving developing states of their responsibilities to work toward development for their citizens at home. Unlike the experts of the Civil Society process, the experts of the IAMR were drawn from those with mandates from self-organized, grassroots migrant groups.

Bustamante's participation in the IAMR was interesting because of his official UN role. Indeed, he repeatedly clarified that his participation in the IAMR was as a sociology professor and not in his official capacity. His participation, nevertheless, had the effect, as intended by IAMR organizers, of further exposing the ways in which migrant advocates, including the UN's own designated representative for migrants' concerns, were excluded from the Civil Society processes since he was not officially invited to participate in them.

Academic Knowledge and the Authorization of "Civil Society": Concluding Thoughts

The civil society groups that constitute today's transnational social movements include self-organized membership-based organizations as well as NGOs. Often, however, scholars fail to make the distinctions between these types of groupings clear. The relative privilege and disadvantage that these distinctions create for groups' ability to engage in transnational struggles is therefore obscured. Indeed, sometimes it is unclear whether we fully understand these distinctions and their implications.

Those of us who are engaged professionally in the work of research and writing, and who are interested in documenting different forms of resistance to neoliberal globalization, but who rely too heavily on the Internet to locate antiglobalization actors, are likely to miss the work of grassroots, self-organized groups. We can be too readily taken in by the technological savvy of NGOs who can occupy significant space in the virtual, if not always in the sensate world. In the case of scholarship on transnational migrants' political mobilizations, it can be all the more difficult to track the activism of the self-organized in the virtual world. Migrants, particularly low-wage workers, as foreigners in their countries of employment may be unable to form formal organizations or apply for funding for their projects because they are not well-enough networked, or worse, because of their legal status as temporary or undocumented workers. There is a real class bias to virtuality that requires the basic hardware to access the Internet (which comes with costs) and the knowledge to write (in English) or time to post material.

Unlike other activists, working-class migrants can be incredibly immobile in spite of the transnationality of their daily existence. Without close attention to these distinctions, which can be due in part to an overreliance on virtual space for research, this oversight can be politically dangerous. These are not simply methodological concerns, but have some significant political implications as we may, wittingly or unwittingly, confer a kind of authority on groups who do not necessarily represent the broader publics on whose behalf they claim to act. In so doing we may give license to governments to mobilize these groups in ways that legitimize states' imperatives. As Nancy Naples (2009) argues, "[W]e must hold the academy responsible for producing knowledge that reveals the complex relations of ruling that contour everyday life" (p. 17). To that end, it becomes necessary for scholars to be attentive to the limits and possibilities of different forms of transnational activism and to be cautious about overly celebratory readings of these activisms. Indeed, it requires that we engage research that allows us insight into local contexts of struggle as it is in those sites where we might identify forms of resistance not immediately apparent in transnational spaces (virtual and otherwise).

I make this point because the IAMR, for instance, would have been nearly invisible to scholars, because the groups that convened it do not have the same kind of visibility on the Internet as their counterparts which organized the PGAMDHR event, and were ultimately involved in the official GFMD Civil Society meetings. Their inability to participate in these meetings has to do, in part, with the fact that they are often less internationally networked and less visible on the Internet than their wealthier counterparts. It is also, however, a consequence of very different orientations to social change. The IAMR conveners prioritize the painstaking and grueling work of social movement building, which depends on face-to-face interactions for organizational growth. The Internet may be a tool to sustain interactions between social movement actors, but it is often less important than cheaper and more accessible communication technologies like cellular and landline telephones. The Internet is not a primary mode of organizing. This is not to say that the IAMR conveners are not interested in making transnational linkages. The IAMR was organized precisely because its conveners are invested in them, yet their focus on social movement building on a more local level often leaves them out of different kinds of transnational networks populated by professional NGO activists. Indeed, IAMR's conveners, many of whom are inspired by radical anticapitalist politics, are actively excluded from these networks and indeed from sanctioned transnational political spaces.

I hope that we as scholars can instead act as interlocutors for specific groups of self-organized migrants who are actively engaged in knowledge production, but in less formal and accessible ways. Often, as is the case for many of the groups that comprise IMA, grassroots organizations are staffed mainly by volunteers who focus their limited time and resources on the work of organizing rather than on Web updates. The Asian Migrants Coordinating Body, one of the lead organizations behind the IAMR, for instance, has managed to organize tens of thousands of women for demonstrations against the Hong Kong government's efforts to decrease the minimum wage for domestic workers, but one would be hard-pressed to find up-to-the-minute information on the group online. Indeed, it does not even have its own Web site. This is equally true for Migrante-International, a transnational organization of Philippine migrant workers with a global scope, and yet is only thinly represented online.

For those of us who are committed to locating alternative imaginings of a more just social order, it becomes vital to pay attention to the knowledge production of those excluded from official venues and who cannot participate in the circuits, virtual and otherwise, through which others in the "global justice movement" traverse. In order to be able to document the kinds of struggles engaged in by migrant worker activists like Elvira Arellano or Eni Lestari requires some level of political investment and engagement on our part as scholars, for it is in spaces outside of the seats of power, like the space of the street, where migrants can come together not only to narrate their experiences, but also to articulate radical alternatives to the contemporary global order. To share the knowledge of grassroots migrants through our academic writing as well as through other venues, which becomes only possible through our involvement with them, I believe, is therefore a vital task. Grassroots migrants often challenge us to expand our political imaginaries to demand not simply more rights within the existing paradigms of citizenship, but indeed, rights and citizenship of an entirely different order.

Notes

1. See http://www.un.org/esa/population/migration/hld/index.html.
2. http://www.gfmd2008.org/welcome.html
3. http://www.gfmd2008.org/welcome.html
4. http://www.ayala-group.com/
5. http://www.bpi.com.ph/
6. http://site.globe.com.ph/web/guest/home?sid=dtklllr9se02b1254594701474
7. http://www.hoovers.com/sgv-&-co./—ID__153864—/free-co-factsheet.xhtml
8. http://www.magsaysaycareers.com/mco/homepage.aspx
9. http://www.thedclgroup.com/DCLContPlan.html
10. http://www.ercof.org
11. See Khagram, Riker, and Sikkink (2002) for a good distinction between NGOs and grassroots movements.
12. http://www.mfasia.org/peoplesglobalaction/PGAdocuments/html
13. Statement of the International Assembly of Migrants and Refugees, October 25, 2008.

References

Alvarez, S. (1999). Advocating feminism: The Latin American feminist NGO "boom." *International Feminist Journal of Politics* 1(2), 181–209.

Asia Pacific Mission for Migrants (APMM) and International League of People's Struggle Study Commission No. 16. The Global Forum on Migration and Development (GFMD) (October 24, 2008). Development not for the grassroots migrants. p. 7.

della Porta, D., M. Andretta, L. Mosca, & H. Reiter. (2006). *Globalization from below: Transnational activists and protest networks*. Minneapolis: University of Minnesota Press.

Government of Bangladesh and Government of Canada (2008). *Fostering more opportunities for regular migration*. Global Forum on Migration and Development, October 27–30, Manila, Philippines. Retrieved from http://www.gfmd2008.org.

Government of Philippines and Government of United Arab Emirates (2008). *Protecting the rights of migrants—A shared responsibility*. Global Forum on Migration and Development, October 27–30, Manila, Philippines. Retrieved from http://www.gfmd2008.org.

Jad, I. (2007). NGOs: Between buzzwords and social movements. *Development in Practice* 17(4–5): 624–629.

Khagram, S., J. Riker, & K. Sikkink. (2002). From Santiago to Seattle: Transnational advocacy groups restructuring world politics. In *Restructuring world politics: Transnational social movements, networks and norms*, ed. S., Khagram, J., Riker, & K. Sikkink, pp. 3–23. Minneapolis: University of Minnesota Press.

Naples, N. (2009). Crossing borders: Community activism, globalization and social justice. *Social Problems* 56(1): 3–20.

Rodriguez, R. M. (2010). *Migrants for export: How the Philippine state brokers labor to the world*. Minneapolis: University of Minnesota Press.

Santo Tomas, P. (2008). *Protecting migrant workers: A shared responsibility*. Roundtable paper presentation for GFMD Civil Society Days, October 27–28. Manila, Philippines. Retrieved from http://www.gfmd2008.org/conference-documents.html.

Sharma, N. (2005). Anti-trafficking rhetoric and the making of global apartheid. *National Women's Studies Association Journal* 17(3): 88–111.

Ziai, A. & Schwenken, H. (2009). *The governance of migration as development policy?* Unpublished manuscript. The International Organization for Migration.

CHAPTER 5

Whatever Happened to the Counter-Globalization Movement? Some Reflections on Antagonism, Vanguardism, and Professionalization

Kees Hudig and Emma Dowling

"Anarchy in London!" cries the headline on the newsstand outside London Liverpool Street station. The day's papers depict protesters hurling a computer monitor and keyboard through a smashed window of a bank. Amid the shattered glass of spectacle, an engulfing sea of photographers stands eager to get a shot. At the G20 protests in April 2009, there seemed to be more media than protesters, more tweeting than blockading, more camera flashes than flares. The disjuncture of protest ritual and media hype in light of an absent social force was evident as we walked away from the Bank of England blockade. We felt a sense of irony, but also frustration. The economic system faced a crisis big enough to make those in power fear global capitalist meltdown, but there was no more than a glimpse left of the powerful anticapitalist movement evolving around the turn of the millennium. This global "movement of movements" had been a harbinger of the current crisis, with its message of the massive injustices and evident instabilities of neoliberal globalization. Against the story of neoliberalism's inevitability, there had been another. Yet the dreams of other possible worlds now appeared even more fanciful than the movement's critics had reported at the time.

According to some social movement theorists, movements "decline" when their "message" becomes distilled into "common sense" (Tarrow, 2000; Zolberg, 1972), having shifted the grounds of the possible. And yet, as the state ushers in more privatization, more free trade, and more financialization as the only solutions to the current crisis, it is clear that the existing bottom line, precisely not the undesirability of neoliberalism, but the ideology of the lack of any "viable" alternative, still looms large. But where has the counter-globalization movement[1] got to?

Making Sense of the Current Impasse

Our reflections here are situated within our sense of a current movement impasse. Struggling for a way out, we pause to identify and make sense of what we can learn from our personal-political experiences of organizing. This we do as a form of knowledge production that we understand to be an integral part of the political practice of social movement activism. Such a form of knowledge production takes as its starting point concrete experiences in thinking through the frustrations encountered, formulating these as sites of inquiry that can reveal insights into possible steps forward. In this chapter, we turn our attention to the vulnerability of horizontal self-organized networks. To assess our frustrations in terms of their challenges to horizontal self-organized networks, we draw on examples from the Social Forum process and summit protests in Europe. In order to contextualize our analysis, we first provide a sketch of the development of this global movement.

The Explosion of Movement: Movements of Networks or Networks of Movements?

The millennium ended with a surprising, powerful resurrection[2] of massive and often militant mobilizations in the Global North. Following struggles against the International Monetary Fund (IMF) and Structural Adjustment Policies (SAPs) across the developing world in the mid 1980s, protests against global summits gathered momentum.[3] People's Global Action (PGA)—blossoming from the Zapatista *encuentros*—developed a model of decentralized yet simultaneous action that motivated thousands of people to confront global capitalism in a wave of uncompromising as well as carnivalesque street actions. 1999 was especially exciting, with the anti-G8 and anti-EU summit protests in Cologne, Germany and the "Carnival Against Capital" in London, UK.[4] Then, in November, the WTO protests in Seattle happened. This event resonated massively across global mass media and took local authorities by surprise. A movement that had already been boiling in the global South was clearly emerging in the North, with global (or at least transnational) dimensions.

This new transnational movement had no clear center, no visible leadership or any formal organizational structures. It consisted of networks and was not *a* movement, but rather a *movement of movements* (Klein, 2001). This was less premeditated than it was the result of an organic process. Following the neoliberal attacks on the organized left of the Thatcher and Reagan years, the 1990s had left the global North a political desert. Most existing structures had either been dismantled or had shifted to the center. The political party was losing appeal as an instrument of social change because of the inabilities of parliamentary democracy to command the tides of neoliberal globalization. On the level of social movement organizing, the monolithic "movement" became a thing of the past and contrasted with what were considered more transparent and decentralized—"horizontal"— forms of organizing.[5] Nonhierarchical and grassroots-orientated forms of organizing had existed since the 1960s, but what was novel was their much broader dissemination, along with the privileging of organically developing "networks," considered more cooperative, more open to diversity, and without the need for political programs or leaders.

After Seattle, these networks gained incredible momentum, described as the "Seattle effect" (Juris, 2008). Marches and militant actions appeared almost anywhere politicians and corporate leaders tried to meet. In 2000, the World Economic Forum (WEF) in Davos, a secluded Swiss ski resort, was met by angry demonstrations and riots. Tens of thousands of demonstrators blockaded the spring meeting of the World Bank and IMF in Washington, D.C. Later that year, Prague was the scene of massive confrontational demonstrations against the World Bank and IMF, as was Nice when it hosted the European Union summit. 2001 looked like a yearlong political battle zone with huge mobilizations and riots against the Summit of the Americas in Quebec, the EU summit in Gothenburg, and the G8 in Genoa.[6]

Turning Point: Porto Alegre's World Social Forum

The idea to organize a World Social Forum (WSF) as a counter-summit to the WEF emerged after demonstrations in Davos in 2000 once again ended in clashes. Key initiators of the social forums were figures from the originally French organization ATTAC with other left-wing personalities, mainly from older, previously existing "vertical" political traditions and cultures (Nunes, 2005). The idea was to create a space where movement representatives could discuss alternatives to neoliberal globalization, thus moving beyond mere protest against the system. Whether it was explicitly planned or not, the organization of the WSF and other local forums meant that direct conflict with the ruling system waned somewhat. Forums took place away from the official summits, to a certain extent playing off direct confrontation with the development of alternatives and setting up a false dichotomy at the expense of an antagonistic politics. At the same time, social forums have proved to be valuable spaces for debate, exchange, and networking, much beyond anything the original organizers may have envisaged, and which it would be myopic to ignore.[7] An inspiring model of direct democracy, social forums captured the imagination of many. They were shaped by the organizational ideal of broad and diverse networks with no central political decision-making body, and a charter of principles[8] that put nonrepresentational politics and process center stage, along with an ethos of the forum as an open space (see also http://www.openspaceforum.net). It was also the only way the WSF could have happened, since nobody could claim "ownership" of this "movement," and many active inside it cherished the horizontal, undogmatic, and open model.

What had emerged was an enormous breadth of global social justice activism; the spectrum ranged from social democratic and pacifist church groups to militant anarchists, single-issue movements, and different left-wing tendencies. This movement encompassed an intermeshing of new networks as well as different past ways of organizing, from feminist and autonomous struggles of the 1960s and 1970s, as well as more vertical forms of organizing that were part of the repertoire of "older" social movements such as trade unions and political parties. Participants worked on a basis of "agreeing to disagree" where necessary, and, most important, while in the process of working together, could find new ways of doing so productively. Precisely through negotiating the diversity of subjectivities, political positions, as well as preferred tactics, movements could learn together and forge a new common politics in this process of—often necessarily conflictual—engagement. For many, emphasis lay on the forging of new social relations, as opposed to formulating political demands on the basis of existing identities or subject positions, often perceived as a form of organizing from a previous era. Below, we reflect upon some of our experiences of these processes in Europe.

Building Movement—The Politics of Negotiating Space and Tactics

Summit protests were also conflictual spaces in terms of the movement's internal politics. One example here is the issue of "violence," or to what extent it was acceptable or productive to negotiate agreements with the authorities.⁹ This conflict led to different political-tactical agreements on how and whether to share common spaces, often reflecting the power dynamics within a particular mobilization at a given time. For example, in Prague 2000, one of the first real rifts emerged, provoking the emergence of the *pink and silver* current at protests. Protesters had been divided up into three separate color-coded groups that were to approach the summit venue from different directions. The more carnivalesque British Reclaim the Streets (RTS) activists with their "tactical frivolity" and samba bands were thrown together with the Socialist Workers Party (SWP), who are from the UK and known for entering and attempting to dominate campaigns and recruit members, and their international family of Trotskyists. They were all supposed to be pink, but to differentiate themselves, RTS chose pink and silver. Before the blockades started, the Trotskyist section—who apparently had reservations about their members being involved in direct action—withdrew, saying they had planned an international strategy meeting elsewhere in the city. Many activist groups refused to work together with them afterward. Even more fractious were the anti-G8 protests in Gleneagles in 2005, where three almost completely separate mobilizations took place between the different strands of the movement, with minimal and reluctant sharing of space and resources. In Heiligendamm in 2007, on the other hand, there was a much more coherent political process in which different movement actors tried to negotiate a common action consensus for the blockades (Block G8), which forged common ground and fostered more channels of communication for negotiation and common action with ultimately more possibilities for movement building.

One long-standing point of contention has been the concurrent holding of counter-summits versus blockading actions. The preferred action form of more moderate NGOs was counter-summits, because they are less confrontational.¹⁰ Counter-summits consisted of workshops and meetings at which participants could network and exchange information about political issues and campaigns. More radical activists preferred to blockade and organize forms of civil disobedience and direct action against the summit, arguing that the counter-summit should not be held at the same time in order not to distract from the protests, while counter-summit proponents would argue that it was precisely by holding the counter-summit at the same time as the summit that public awareness of alternative voices and visions could be generated. In many cases both sides pursued their own agendas, thus weakening the potential common force that could have been forged.

Self-managed protest camps were further spaces where knowledge, skill, and resource sharing took place. Activists from different backgrounds and political currents could all play their part, but this also meant that different actors coming from very different traditions could organize in their own way with little compromise, organizing almost in parallel (Maeckelbergh, 2009). These protest camps were mostly set up and run by autonomous horizontal direct-action activists, most of them young, building upon knowledge and experiences stemming from antinuclear, environmental, and squatting movements that had existed since the 1970s. On the whole, neither vanguard political parties nor the larger NGOs were involved in maintaining this infrastructure, but nonetheless tried to mobilize "masses" to join the demonstrations (if not always the more militant actions), thus attempting to occupy the public arena, while the work of building infrastructure was dominated by autonomous horizontal direct-action activists. Importantly, the "horizontals" understood this work as immensely important and political, forming a vital part of a prefigurative[11] politics (Maeckelbergh, 2009), namely building alternative reproductive structures in the present, not at some future date after power had been seized.

Power Struggles within Social Forums

Our experience of social forums in Europe has shown some disadvantages of the horizontal network model. In the context of a political economy of social forums, the vulnerabilities pertain to being overly influenced by (local) governments, professionalized NGOs, and political parties. For example, the effects of the presence of disguised vanguardism could be clearly seen at the first European Social Forum (ESF) in 2002 in Florence, Italy. With the invasion of Iraq looming, the SWP had decided that all of its members had to spread out over the hundreds of workshops and seminars to promote the idea of bending the focus of the counter-globalization movement toward an antiwar focus. The SWP launches campaigns like corporations advertise a new product. Local branches adopt one "simple" theme as the central one (e.g., stopping the war) out of conviction that ordinary people will not be convinced by complex stories about different and interrelated issues. Other issues had to make way. This is one of the reasons that issues like free trade, mining and natural resources, and the environment disappeared from the agenda.

The organization of the third ESF in London in 2004, was a particularly fraught process marked by major conflicts over the control of the event. On the one hand, there was a strategic alliance between political forces who were to become dubbed by the self-declared "horizontals" as the "verticals," an alliance of predominantly Trotskyist or allied political forces from political parties, some trade unions, and the London mayor's office; the "horizontals" constituted a mixed bag of anarchists, autonomists, media activists, grassroots NGOs, and feminist and human rights groups.[12] During the process, the Trotskyist groups—with the support of the mayor's office—succeeded in taking over and controlling the official organizing process through intimidation, stacking meetings, and using access to resources as blackmail in negotiations in order to produce a centralized, politically uncontroversial talking-heads event from which to recruit members for their organizations. This was an arduous process with severe political consequences in terms of how the event was emptied of much of its emancipatory potential, but it was also an excellent example of the significance of the dynamics of knowledge production within social movements. The direct confrontation between different movement actors with different ideas about the means as well as the ends of social movement activism led to the development of new terms that function as mechanisms through which to negotiate the desired form and direction of social change.

The charter of principles that the WSF uses relies heavily on political goodwill. The difficulties of establishing and maintaining such a political culture of consensus decision making, openness, and transparency in the organizing process are further exacerbated by material realities: The vulnerability of the forum becomes subject to the political economy of access to resources. In addition, when activists and movement leaders detach themselves from ordinary people and perceive themselves as "professionals," the social forums become less accessible for nonactivists and they become spaces of alienation rather than spaces of empowerment. This sense of professionalism, along with vanguardist notions of centralization and the necessary seizure of governmental power, can also lead to the deferral of the infamous "other possible world" to some point in the future, instead of understanding it to be located within the present and part and parcel of any organizing process. In this way political goals are relegated to a level of abstraction in which oppressive and exploitative practices of capitalism, racism, sexism, or imperialism are criticized, yet the link to one's own political practices, behaviors, and silencings in the present are not made (see also Dowling, 2005).

During the 2004 ESF organizing process, at the point of impasse, the only option became exodus; a retreat from the official forum into the self-organized spaces, autonomous from the official forum. These built upon and reinforced the existing tradition of more radical libertarian groups that had been operating on a position of "one foot inside, one foot out" since the first ESF in Florence in 2002.[13] Many participants and speakers would act in both arenas, while "radicals" could organize alternative events without frontal opposition to the official ESF, although it proved to be difficult to "compete" with the official program without access to the same means and funding. However, this also meant that there was more space to actually use the self-organized infrastructures and knowledge of the movement (open source software, kitchen collectives, consensus-decision making, free entry). This also built upon the same kind of political practices and initiatives as the Intercontinental Youth Camps of the WSF (see Nunes, 2005).

The Dutch political landscape is perhaps one of the most "managed" in the world. Its much-hyped "consensus" model draws many critical forces toward the center and exercises "friendly exclusion" toward those who try to remain radical. Almost all organizations that constituted critical forces in the 1960s later mutated into professional NGOs with a certain degree of access to decision making on all levels of government, losing any active membership base. Currently, any relationship with a membership consists mainly of collecting membership fees that make it possible for staff to campaign.[14] Increasingly, these organizations are funded by the Dutch government. To counter growing opposition from the right to the government supposedly "wasting taxpayers' money on self-created opposition," proof was required that the NGOs still have a real base. Social forums offered the NGOs an opportunity to provide this evidence. NGOs had money to fund the forums and thus became central players in the organizing process. Interestingly, one could also see the manufacturing of a need for funding by making the forum far more expensive than necessary. Instead of organizing the forums in self-managed spaces and noncommercialized structures, the project was immediately framed in terms of renting "representative places" and paying for services. The justification was that to reach beyond the "usual activist circles," a "professional" environment was required. The three Dutch Social Forums (Amsterdam 2005, Nijmegen 2006, and Amsterdam 2007 during the G8 Summit) were mainly organized by a coalition of professional NGOs, trade unions, and the local Trotskyist organization Internationale Socialisten (International Socialists). The Dutch Social Forum process took off comparatively late due to the reluctance of grassroots organizations to get involved, precisely because they foresaw where it was headed. It started in 2004 with a general open meeting, where tensions arose when professional NGO and trade union staff were accused of manipulating the process. Grassroots groups started up their own public mailing list to discuss this (see https://lists.riseup.net/www/arc/transparant.nsf). From then on, relations between so-called professionals and grassrooters deteriorated. Initially there were attempts to keep the two worlds together, and the grassrooters were offered seats on the "executive board" of the Dutch Social Forum. Many of the grassrooters questioned the need for such a nontransparent "board of delegates" and criticized the process, accusing the board of not executing decisions taken at the general meetings. The fact that the board refused to minute meetings did not help address these kinds of accusations. When voting was suggested in the case of difficult decisions, some participants dismissed this as a return to the bad practices of an old world. The Dutch forum is possibly the only social forum to replace the original WSF charter of principles with a completely meaningless one that omitted the radical descriptions contained in the original charter. The logic was that the revised charter would make it possible to include organizations (mainly large NGOs and trade unions) not opposed to neoliberalism. Most radical and grassroots organizations quit the project in the run-up to this forum, and felt confirmed in their decision when they heard that the program contained workshops that were seriously discussing the merits of neoliberalism. Nonetheless, some critics still organized part of the program within the forum (and were not prevented from doing so). After a relatively unsuccessful attempt to imitate the UK "Make Poverty History" model in the Netherlands in the run-up to the 2007 G8 summit in Germany, the big NGOs and trade unions decided to withdraw funding and the few remaining executives decided that the Dutch Social Forum should go into "hibernation."

Professionalization of Activism and the Role of NGOs

Lately there has been an increase in attention to the dynamics between professional NGOs and grassroots organizations (Barker, 2007; INCITE! Women of Color Against Violence, 2007). For example, Barker (2007) develops the democracy manipulation model based upon Herman and Chomsky's (1988) media manipulation model, which explained how manipulation functions in democratic capitalist conditions where no official censorship exists. Barker argues that while liberal funds and NGOs do not set out to manipulate, they—often unconsciously—steer the fields in which they operate toward a certain direction. In particular, Barker points to the problem of professionalization. While one cannot speak of any bad intentions, the result is not apolitical; the modus operandi deployed functions to exclude those participants who cannot or do not want to operate in certain professionalized—and alienated—ways. Similar to the media filters that Herman and Chomsky identified, Barker (2008) identifies kinds of professionalized filters that are seldom acknowledged by NGO personnel when they operate in networks like the social forum. Moreover, NGOs are careful about their relations with governments or individual donors who do not accept antagonistic politics. They tend to move away from confrontational politics and try to influence the mainstream political agenda, steering collective campaigns away from illegal actions, arrests, and confrontations with state power. This means constant pressure to move activities away from direct contestation and toward more symbolic actions that do not involve challenging power-holders directly or in any material way, deploying parliamentary strategies that remain within existing boundaries of political legitimacy. One example is the recent "stand up against poverty" campaign, part of the Global Call to Action Against Poverty (G-CAP) initiatives in which at a certain time of day, people collectively meet and literally stand up to signal that they are against poverty, where poverty is constructed as an abstract entity as opposed to a social and economic relationship of power within which social and political subjects are embedded.

This problem of subjectivity, of organizing on behalf of one's own interests vis-à-vis capitalist relations, and organizing on behalf of a cause permeates these political processes and can lead to completely different agendas. Not least, the reality of political economy also creates unequal influence when the staff of big NGOs have the money to fly around the world and network, preparing for the next forums as part of their paid jobs, while members of grassroots movements often have less access to resources and have to do their political work in their spare time when they are not engaged in waged work.

The best—or worst—example of the development of NGO practices was the Make Poverty History (MPH) coalition during the 2005 UK Gleneagles G8 summit. Led by British poverty and development NGOs such as Oxfam, Christian Aid, Action Aid, and CAFOD, the Make Poverty History Campaign brought together a broad range of charitable organizations, church groups, local councils, members of the public, businesses, and celebrities; that is, "civil society." The politics were defined by the choice to lobby the G8 (via the British government) to implement policies that would alleviate extreme poverty in developing countries through trade justice, debt cancellation, and "more and better aid." The campaign began in the autumn of 2004 and involved raising awareness to mobilize broad sections of the British public to support the campaign and its three demands. Coalition partners would sell white wristbands symbolizing people's support for the initiative and thus raise money for their individual projects. Celebrities were brought on board with TV advertisements. Given the propensity toward celebrity culture in Britain, this was a successful strategy in terms of appealing to public support. MPH fostered close links with the British government, which resulted in the campaign actively supporting government and the G8 summit's relegitimation strategies instead of contesting them. MPH's emphasis was very much on putting moral pressure on the G8 by appearing as civilized and peaceful, representative of broad sections of the British public, and in opposition to the more antagonistic counter-globalization movement. This strategy is indicative of a general trend toward professionalization of dissent; turning away from general, often spontaneous resistance and protest, and turning toward professional campaigning, often of the single-issue kind, with "clear demands" toward the political class for solutions, and itself becoming subsumed under the logic and requisites of capital even beyond the constraints of state funding toward the development of the business model for social movements (see for example, Sireau, 2009, for this kind of approach).

But where do the political acumen, the vibrancy, and innovation for such social movement business models come from? It was not a coincidence that the anti-globalization movement was called a Do It Yourself movement, with "trial and error" as one of the main principles. Whether with media, street actions, or tactics to outsmart police, there was constant invention and renewal in these radical activist circles, from Indymedia to banner dropping, blocking streets, or running spoof advertising campaigns parodying multinationals. Many large corporate-structured NGOs look closely at successful radical methods and campaigns, then copy the methods that seem "cool" or attractive to young people (or employ those same activists who had come up with them), albeit applying them in a "light" and inoffensive version and so making them "toothless." During Prague, for the first time a popular band tried to mobilize their fans to join the protest against the IMF/World Bank. Soon after, Oxfam International hired another band to advertise its poverty reduction campaigning at the WTO summit in Cancun. Following the logic of systematization, Oxfam advised its local chapters to do the same with a local pop group. In the Netherlands, Novib, the development NGO and member of Oxfam, signed a contract with the band Krezip, whose front woman disclosed in the magazine *Onze Wereld* (*Our World*) that her band spent some time negotiating the campaign they would help advertise "as we didn't want to be forced not to wear corporate sneakers if we liked them,"(NRC Handelsblad, 2005) a statement that points to the level of political alienation that occurs with such forms of professionalization.

The Other Side of the Coin: Problems with Horizontal Radicalism

It seems obvious to us that radical groups have not yet found strategies to counter their remarginalization. One problem here is the huge opposition to political reflection and analysis, with the emphasis on "doing" as opposed to "theorizing," which draws false and counterproductive dichotomies between these two important and interconnected modes of activity. Ironically, this has gone hand in hand with difficulties in strategic planning and the reaching of agreements over longer periods of time, making it difficult to find more long-term solutions. Being "unprofessional" (or perhaps "amateurs," in the true meaning of the word: doing things because you like to do them and not because you are getting paid for it) also has serious disadvantages, especially if you have to work in a difficult network with other participants. Grassroots movements, especially more radical ones, tend to be unstable and are not necessarily more transparent than NGOs or vanguard parties given the existence of unspoken or invisible hierarchies, or simply lack of access to anyone who is not already in an inner circle. Radical groups can be unreliable in their focus, which can change rapidly given changing interests. Moreover, agendas can be greatly influenced by such factors as charismatic leadership. Some groups are uncompromising on militant tactics and are unwilling or unable to make agreements with less militant organizations, which makes collaboration difficult. They will do what they want at any time, which can result in being both dogmatic and manipulative in negotiations with other groups. It can be difficult to keep these forces together in the same campaign, which weakens the movement overall.

A further such weakening has been the increasing separation and entrenchment of the different spaces at the expense of fruitful cross-pollination of ideas and ideologies. For example, at the ESF in Malmo in 2008, the autonomous space was so remote and badly publicized that most ESF participants never found it. Maintaining their safe space against the odds of insufficient funds and excessive police attention and intimidation also meant that the "horizontals" for the most part did not visit and contribute politically to the official ESF. The different dynamics of these spaces are evident in the remarks of one campaigner who presented in both spaces:

> At both workshops (about the upcoming protests against the Strasbourg NATO summit) there were some sixty participants. At the ESF one, the participants were only interested in adding information to the introduction I had given. Nobody seemed interested in the actual protests, it could easily have been something at a university. At the autonomous space very little was about the background of NATO as such, people were eager to know details about the protests and the groups present immediately began to make plans to for the protests.

The idea that these two worlds should mingle still seems a good one. However, the gap between these different worlds or mode of doing politics has widened, not closed.

Moving out of the Impasse?

What was so exciting about the counter-globalization movement were the potential seeds that were sown for ordinary everyday people to come together and network transnationally against the injustices of a globalized system and organize alternative ways of reproducing their common livelihoods. Summit protests and social forums were both immensely important in creating the spaces and structures for that to happen. Yet both had their limits. The professionalization of the protest industry and the subsumption under the logic of both the business and the lobby model have been extremely significant and have reduced antagonism. Vanguard parties have also played a role in stifling social movement activism through centralization and control mechanisms that emphasize the event of taking power over the organization of alternatives in the present. At the same time, decentralized grassroots organizing is and always has been a fragile process. We suggest that NGOs be open to criticism with regard to their role as "missionaries" for the state and capital. Moreover, Marxist political parties should disclose their political strategies and be ready to be challenged as opposed to hiding behind front groups in networks. Horizontal grassroots groups should work on building stable structures that are sustainable beyond the immediate subcultural environment and that can both resonate with and involve ordinary people. As the future comes crashing down into the present with the financial crisis, we are reminded of something that the organizers of the Intercontinental Youth Camp, the autonomous space of the WSF, wrote in 2003: "Other worlds are not only possible, but are being bred today. The present is pregnant with the future, and the future lasts a long time" (cited in Nunes, 2005, p. 290).

Notes

1. Movement actors have argued that the struggle is for is a different globalization to the neoliberal kind, thus preferring alter- or counter-globalization over the earlier anti-globalization. We prefer counter-globalization because it encompasses opposition to neoliberal globalization and the desire for an alternative form of globalization.
2. After the fall of the Berlin Wall it seemed that social movements died down in the neoliberal 1990s.
3. For archives of protests, see http://www.gipfelsoli.org; http://www.nadir.org/nadir/initiativ/agp/free/index.html; and http://www.thirdworldtraveler.com/Globalization/Brief_Hx_StrucAdj_DGE.html.
4. There were also actions on several other continents and in forty-three different countries. For example, in Millau, France, the farmers' organization *Confédération Paysanne* dismantled a McDonald's restaurant; solidarity actions followed the arrests of nine farmers.

5. This term emerged during the Argentinean insurgence in December 2001, where the main slogan was "Que se Vayan Todos" ("Let them—all politicians—go to hell"), which resonated in many places worldwide.
6. The flip side of this has been a kind of "copy/paste" form of organizing: Tactics are copied with little consideration for the specificity of locality, meaning the "effect" remained momentary because it was not rooted in local, ongoing social movement activism.
7. The Indian organization Research Unit for Political Economy (RUPE) (2003) calls this "globalization with a human face," a "third way." Petras (2002) has criticized the WSF as a platform for moderate politicians that keeps radical movements out. Famously, the EZLN is actually formally excluded from the WSF (Sen, 2005). For a geographic overview of social forums, see Anheier, Glasius, and Kaldor (2005).
8. WSF Charter of Principles, http://www.forumsocialmundial.org.br/main.php?id_menu=4&cd_language=2.
9. Divergent positions regarding confrontational tactics were a source of contention and influenced the development of social forums as an alternative form of convergence.
10. The "Other Economic Summit" (TOES), a counter-summit organized by the New Economics Foundation at the G7 Summit in 1984 in London, was the first anti-neoliberal summit protest.
11. Prefiguration is a political practice that does not distinguish between the "ultimate values of an ideal society within the very means of struggle for that society" (Maeckelbergh, 2009, p. 67). This explains why, for example, direct democracy and alternative economic practices have been important in counter-globalization movements.
12. Some groups and individuals tried to position themselves in between this dichotomy, either as self-appointed mediators (diagonals), or simply as nonpartisans.
13. In Florence (2002) this was the HUB, in Paris (2003) this was GLAD, in London (2004) there were many different autonomous spaces, including Beyond ESF. In Athens this was called ESF Playground.
14. So-called tennermembers are named after the ten euro membership fee.

References

Anheier, H., M. Kaldor, & M. Glasius. (2005). *Global civil society yearbook 2004/5*. Oxford: Oxford University Press.

Barker, M. (2007). *Do capitalists fund revolutions?* Retrieved from http://www.zmag.org/znet/viewArticle/14574.

———. (2008). The Soros media "empire": The power of philanthropy to engineer consent. Retrieved from http://www.swans.com/library/art14/barker02.html.

Dowling, E. (2005). The ethics of engagement revisited: Remembering the ESF 2004. *Ephemera: Theory and Politics in Organisation* 5(2), 205–215.

Herman, E. S., & N. Chomsky. (1988). *Manufacturing consent: The political economy of the mass media*. New York: Pantheon Books.

INCITE! Women of Colour Against Violence (2007). *The revolution will not be funded: Beyond the non-profit industrial complex*. Cambridge, MA: South End Press.

Juris, J. (2008). *Networking futures: The movements against corporate globalisation*. Durham NC: Duke University Press.

Klein, N. (2001). Reclaiming the commons. *New Left Review 9*, 81–89.

Maeckelbergh, M. (2009). *The will of the many: How the alterglobalisation movement is changing the face of democracy*. London: Pluto.

NRC Handelsblad (November 19, 2005). Feestend Verbeteren Wij de Wereld [Partying we work on a better world]. Retrieved from http://www.nrc.nl/krant/article1643417.ece/Feestend_verbeteren_wij_de_wereld.

Nunes, R. (2005). The intercontinental youth camp as the unthought of the world social forum. *Ephemera 5*(2), 277–292.

Petras, J. (2002). A tale of two forums. Retrieved from http://www.nadir.org/nadir/initiativ/agp/free/wsf/petras-wsf.htm.

Research Unit for Political Economy [RUPE] (September 2003). Economics and politics of the world social forum. *Aspects of the Indian Economy*, No. 35.

Sen, J. (2005). A tale of two charters. In *The World social forum: Challenging empires*, 1st edition, ed. J. Sen, A. Anand, A. Escobar, & P. Waterman, pp. 72–75. New Delhi: Choike.

Sireau, N. (2009). *Make poverty history: Political communication in action*. Basingstoke: Palgrave Macmillan.

Tarrow, S. (2000). *Power in movement: Social movements and contentious politics*. Cambridge: Cambridge University Press.

Zolberg, A. (1972). Moments of madness. *Politics and Society 2*(2), 183–207.

CHAPTER 6

Collective Approaches to Activist Knowledge: Experiences of the New Anti-Apartheid Movement in Toronto

Rafeef Ziadah and Adam Hanieh

Introduction

The ethnic cleansing of Palestine in 1947–1948 that led to the flight of more than three-quarters of the Palestinian population is not simply a painful historical memory. What Palestinians call *al-Nakba* (catastrophe) remains very much part of lived reality. It is felt in the longing of seven million Palestinians—the world's largest refugee population—to return to their homes and lands from which they were expelled over six decades ago. It is seen in the segregation of more than three million Palestinians in the West Bank and Gaza Strip to scattered population centers divided from one another by Israeli settlements, military checkpoints, and Israeli-only highways. These open-air prisons—surrounded by the "Apartheid Wall"[1] and its associated infrastructure of settlements, military zones, and roads—mean that Palestinians are now confined to approximately twelve percent of historic Palestine. And *al-Nakba* remains with those Palestinians who stayed on their land and became Israeli citizens, forced to live as second-class people in a state built on the destruction of their national identity.

It is sometimes difficult for those living in the West to understand the ongoing reality of *al-Nakba* to the Palestinian existence. The mainstream media presents the question of Palestine as a "conflict," seemingly intractable and fed by an irrational, Byzantine "hate." The solution is seen as coming through the paternalistic intervention of the West—with little regard for the bloody hand of colonialism in the creation of the "Palestine problem" in the first place. Rarely does media coverage step beyond a few short clips from Jerusalem, the West Bank, or Gaza Strip—peppered with reassuring messages of experts and Israeli government spokespeople. The historical reality of Palestinian dispossession—and the fact that Palestinians in the Middle East are now fragmented across at least four different geographical spaces (the Arab world, the West Bank, Gaza Strip, and inside Israel)—is simply lost to the deliberately shallow coverage of mainstream analysis. It is simply not possible to speak of "knowledge" of Palestine without beginning from the understanding that *al-Nakba* is an ongoing process—not a singular point in time.

Yet while dominant academic and popular conceptions have remained mired in an increasingly narrow framework that prevents any real understanding of Palestine, an alternative, counterhegemonic perspective is gaining ground. This perspective has grown from the practical struggle and activity of millions of Palestinians and their supporters across the globe. It is a perspective that acts to resist the fragmentation of the Palestinian people and insists on the ongoing salience of *al-Nakba*, thereby connecting lived history to the contemporary reality within a single narrative. This is the new anti-apartheid movement.

This chapter describes how knowledge has been produced in this movement from our perspectives as organizers in Toronto. Toronto has seen many successes for Palestine solidarity over the past few years. These include being the birthplace of Israeli Apartheid Week—an annual series of campus-based events highlighting the apartheid nature of Israel—in 2005, which has now expanded to sixty different cities across the globe; having the first anti-Israeli apartheid boycott resolution by a Canadian union (the Canadian Union of Public Employees in 2006); and having the first Toronto Palestine Film Festival (attended by more than 4,500 people in 2008). The city has developed a strong and unified Palestine solidarity movement, spanning different campuses, high schools, and unions.

We begin by discussing the basic content of the anti-apartheid perspective and the way in which it emerged as part of a struggle by Palestinians to recenter solidarity on the lived reality of *al-Nakba*. In this sense, the knowledge produced by this movement runs up against dominant perspectives on Palestine—particularly those that were hegemonic through the 1990s. The second part of this chapter discusses the organizational and ideological practices that enabled this knowledge production to take place. We highlight four major themes in this regard: keeping the Palestinian narrative central, nonsectarian movement building, sustainable knowledge sharing and learning, and open participatory structures. We write as two Palestinians involved in this movement but with the understanding that we are just a small part of a much larger global struggle. As such, our perspectives are written as two activists within North America who are simultaneously Palestinian but also part of the solidarity movement. We do not intend or claim to speak on behalf of either.

The Apartheid Analysis: The Growth of a Counterhegemonic Narrative

To gain a sense of the reality of the Palestinian people, we need look no further than the Gaza Strip. Only forty-five kilometers long and about ten to twelve kilometers wide, more than 1.4 million Palestinians are crowded into refugee camps throughout the Strip. The vast majority of Gazans are refugees who originate in the population expulsion from areas in the south of Palestine during 1948.[2] A massive electric fence surrounds much of the Strip, and is guarded by the Israeli military. Israel (with Egyptian government support) prevents movement in and out of the area while subjecting its residents to repeated military incursions, shelling, and house demolitions. This happens as the United States, Israel, and the Palestinian Authority continue to speak of negotiations and a two-state solution—a rhetoric that is defied in reality as Israel continues to dictate what is commonly described as "facts on the ground."

This reality is the end point of the disastrous strategic orientation of the Palestinian national movement throughout the 1990s. It is a course that was solidified with the 1993 Oslo Agreement and the famous handshake between Palestinian president Yasser Arafat and Israeli prime minister Yitzhak Rabin on the White House lawn in the warm embrace of Bill Clinton. The Oslo process reduced the demand for Palestinian self-determination to a "state-building" project on ever-shrinking slivers of land managed by a narrow coterie of Palestinian officials in the West Bank and Gaza Strip.[3] After seven years of this reality, Oslo was shattered with the beginning of the Second Intifada (uprising) in September 2000, a mass revolt of the Palestinian people in the West Bank and Gaza that found echo in protests by Palestinian citizens of Israel, mass demonstrations across the Arab world, and the rebirth of the solidarity movement outside the region.

Before getting to the content of the new solidarity movement, it is important to emphasize that the Oslo process brought disaster not only to Palestinians living in the West Bank and Gaza, but also to wider solidarity efforts. Although Palestinian opposition groups and various intellectuals continuously warned against the Oslo project, the effect it had on the "older" Palestine solidarity movement during the 1990s was devastating. Networks that had developed over decades in the 1960s, 1970s, and 1980s simply fell apart as confusion over the future of Palestine and the Palestinians caused disillusionment among a broad layer of activists. Oslo had acted to demoralize and weaken the earlier structures of Palestinian solidarity.

With the outbreak of the Second Intifada, however, these structures began to revive and regenerate. The movement grew as Israeli violations of Palestinian human rights intensified, particularly following the reinvasion of the West Bank by Israeli troops in March–April 2002. The solidarity movement organized mass demonstrations and public meetings, and some activists began circulating petitions calling for divestment from Israel. These activities dovetailed with the massive demonstrations that erupted at the time of the war against Iraq in 2003. Many of the same activists involved in Palestine solidarity also led the mobilizations against the war in Iraq, drawing the lessons that a perspective on Palestine needed to be grounded in a broader anti-imperialist framework.

A real turning point took place in 2004–2005 when a variety of efforts around the globe began to coalesce around an analysis of Israel as an apartheid state that demanded a strategic response of boycott, divestment, and sanctions in the manner of the struggle against South African apartheid. This vision found an organizing center when the Unified Call for Boycott, Divestment, and Sanctions (BDS) came from Palestine in 2005, signed by more than 170 civil society organizations. The Unified BDS Call was the most authoritative and widely supported strategic statement to have emerged from Palestine in decades. All political factions, labor, student and women's organizations, and refugee groups across the Arab world supported the call.[4]

The global articulation of this movement is an apt illustration of what scholars have described as the growth of a new transnational activism (Colas, 2001; Keck & Sikkink, 1998; Smith, Chatfield, & Pangucco, 1997). Indeed, it is somewhat ironic that a largely stateless people have been at the forefront of a mode of activism that is grounded in a transformation of the relationship between states. The 2005 BDS call acted to link all sectors of the Palestinian people (rather than just focusing on those living in the West Bank and Gaza Strip), and also connected activists at a global level within a single movement. In this sense, the BDS call multiplied the resources and strengths of a range of actors across the international system (Keck & Sikkink, 1998).

The explicit linkage of all sectors of the Palestinian population within the 2005 BDS call was summarized in its three concise goals: the right of return of all Palestinian refugees, an end to the occupation of all Arab lands, and full equality for Palestinian citizens of Israel. In this sense, the BDS call acted against the disastrous trajectory of the Oslo process, its fragmentation of the Palestinian people, and its reduction to simply a bargaining over land in the West Bank and Gaza.

The call also explicitly adopted an analysis of Israel as an apartheid state, which had a galvanizing affect on the solidarity movement. This was not a new perspective—Palestinians and anti-Zionist Israelis had long made the comparisons with South African apartheid (as indeed had leading Zionists and South African apartheid government representatives)[5]—but it was significant in that this discourse had been largely lost during the 1990s. Many North American activists preferred to focus on the crimes of the occupation in the West Bank and Gaza while ignoring the broader systemic questions. These individuals argued that it was necessary to stay within a framework focused on UN resolutions and international law rather than adopt an apartheid analysis, because it would be "too radical."

Yet the exact opposite was true—the apartheid analysis simply made sense to people. It addressed the problem at its roots and motivated people to take action. The most vocal arguments for this strategic perspective came from Palestinian activists, many of them refugees, who pointed out that an analysis that concentrated solely on the occupation in the West Bank and Gaza denied the reality of *al-Nakba* and did not address the self-definition of Israel as an exclusively Jewish state.[6] Israel defines itself as a Jewish state and, therefore, cannot be a state for all of its citizens. The apartheid character has been clear from Israel's inception. It is illustrated by the fact that Palestinian refugees are prevented from returning to their homes and lands from which they were expelled. In contrast, any person of Jewish descent from anywhere in the world may become an Israeli citizen under the so-called Law of Return.[7] The apartheid analysis bridged all parts of the Palestinian people: those who were citizens of Israel, those living in the West Bank and Gaza Strip, and those in exile. It is a strategy grounded in the right of return of Palestinian refugees to their homes and lands. The analytic link with South African apartheid helped to clarify the real nature of Zionism as an exclusivist settler-colonial project. The strategic demands of boycott, divestment, and sanctions helped to illustrate the powerful ties between North American and European capitalism and the Zionist state. We could also build upon the experiences and lessons of the earlier movement against South African apartheid.

Despite the analytical differences, it was important to work with the wing of the Palestine solidarity movement that did consider themselves solely anti-occupation and resisted adopting a broader apartheid framework. It was not a question of putting "a line" forward and expecting everyone else to follow—rather it was an educational process that needed to be engaged in. In Canada, the movement quickly gained momentum from 2005 onward. Fed up with the empty rhetoric and cheap condemnation of the "violence on both sides," people began to feel that they had power to make an impact. The BDS campaign enabled unions, student groups, and cultural and religious organizations to demonstrate a popular refusal to participate in and sustain the structures of racial discrimination and oppression. It helped to break the fragmented and stunted nature of Palestine solidarity work that tended to fall into a never-ending response to Israeli massacres rather than providing an alternative action to stop those massacres. The BDS campaign did not replace the need for outreach, education, and action around the myriad issues connected to Palestine such as refugees, the apartheid wall, or prisoners. Rather, it answered the question: What action do we take after we have understood the nature of Israel as an apartheid state? It provided a concrete strategic focus and plan of action that helped to raise consciousness around Palestine as it was carried out.[8]

By creating an environment where it was widely seen as morally repugnant to be openly associated with Israeli apartheid, the movement helped to undermine the silent acquiescence that Israel had relied upon for decades. The successes in this regard could be felt as writers, artists, filmmakers, and others began to distance themselves from Israel in a manner reminiscent of the South African anti-apartheid struggle. Interestingly, one of the exciting and unexpected consequences of this work was the rebirth of consciousness among a younger generation about the struggles and achievements of the South African anti-apartheid struggle—this was something that had largely dropped from collective memory.

The movement also had a big impact on the left. Overall, the organized political left in North America lagged several steps behind the developments in the movement. Despite having a programmatically correct position on Palestine (condemnation of Zionism, an understanding of Israel's settler-colonial nature), much of the left tail-ended the BDS campaign (and some organizations even argued against it initially, or gave muted public support while attempting to undermine it behind the scenes). The reasons for this relate to the deep-seated belief of much of the organized North American left that its role is to bring the "correct line" to the movement from the outside, rather than understanding that a political line is something forged through engagement with the struggle. In this sense, much of the North American left simply displayed an anti-Marxist perspective.[9]

Having said this, a broader sense of radical politics has been key to building the BDS movement in Toronto. The reason for this is that the question of Palestine is part of a larger struggle that is anticapitalist and anti-imperialist. Most significantly in the Canadian context, this means understanding the linkages with Indigenous struggles and the necessity of supporting the variety of anticolonial struggles. Just as the Palestinian *al-Nakba* is ongoing, so is the genocide against Indigenous Peoples in Canada. Having a broader sense of radical politics has acted to strengthen our movements because it means we are fighting alongside others. Contrary to the liberal perspective that tells us "don't be too radical because you will scare people away," the reality is the opposite. A sole focus on Palestine and refusal to take sides and stand with other struggles actually means to narrow and weaken our movements. By default, it leads to aligning ourselves with those in power—a recipe for disaster.

With this overview of how the movement has grown and developed, we now turn to the organizational and ideological practices that helped foster the development of this counterhegemonic knowledge. We will highlight four aspects of this: keeping the Palestinian narrative central, nonsectarian movement building, sustainable knowledge sharing and learning, and open participatory structures.

Keeping the Palestinian Narrative Central

Feminist and Indigenous scholars have particularly noted the role that knowledge production plays in silencing the voices of the marginalized and reinforcing dominant power structures (Mihesuah & Wilson, 2004; Mohanty, 2003). While much of this literature is intended as a critique of academic disciplines such as anthropology and sociology, we can extend many of the same themes to the nature of knowledge production within activist spaces. Too often, knowledge production within North American solidarity movements tends to ignore the voices of those with whom we are supposed to be acting in solidarity.

One of the basic tenets of the new anti-apartheid movement is to work consciously against this silencing. The Palestinian narrative should be central to all solidarity activities, and the knowledge produced by our movements must be grounded in the history of Palestinians with the voice of the most marginalized at the core—Palestinian refugees. This means that the experience of *al-Nakba* must be kept upfront, with an emphasis on the continued displacement, fragmentation, and dispossession of the Palestinian nation. One example of this was the decision in 2008 to center virtually all Palestine solidarity activities held in Toronto around the theme "Sixty Years of *Nakba*—The Refugees Will Return." This was a conscious strategy adopted through careful consultation with the BDS movement in Palestine. It helped to thread together a variety of different activities and to strengthen consciousness that the Palestinian struggle was not solely a question of the West Bank and Gaza Strip, but centered in a broader historical reality.

Centering the movement around the Palestinian narrative and historical reality produces a different type of knowledge about what needs to be done. Mistakes happen when the lives of people a movement is supposed to be supporting are not seen. This is one reason for the decline of the movement during the 1990s—with the rise of Oslo, the majority of Palestinians were suddenly excluded from the concerns of the movement. Because Zionism is built on the myth of a people without a land for a land without a people (see Weizman, 1983),[10] and because its main project is to render the Palestinian experience invisible, the solidarity movement has to work twice as hard to bring out that experience. This means making sure that Palestinians are allowed to speak for themselves and that their voice is constantly heard.

The flip side to ignoring the Palestinian narrative is its reduction to one of perpetual victimization. To some, often well-meaning individuals, the Palestinian voice is there simply to recount suffering rather than as the key agent of liberation. Knowledge produced in the new anti-apartheid movement must not be based on a tokenizing of Palestinians or projecting an image of victimhood, but rather on solidarity and mutual respect. In Toronto, one way in which this lesson was brought home was through earlier experiences of building solidarity with Palestinian prisoners. The prisoner movement in Palestine is central to the lives of people on the ground, yet prisoners are very clear that they are not victims but people in struggle. This experience with Palestinian prisoners (including several tours of ex-prisoners) helped to reinforce the point that solidarity is best built around notions of struggle, not paternalism and victimhood.

Centering the Palestinian narrative also helps to defy the Zionist myth that Palestinians "never existed." This necessarily means an added emphasis on highlighting Palestinian culture and art as a way to produce knowledge of Palestinian history. One indication of this is the number of Palestinian film festivals that have grown around the world. The first one in Toronto was held in 2008 and was part of the commemoration of *al-Nakba*. Films included documentaries, shorts, features, and comedies. Another cultural avenue is spoken word and hip-hop by Palestinian artists (from both inside and outside Palestine), which helps not only to preserve a

Palestinian identity, but also to connect and push the solidarity movement in interesting ways. Poster art is a third key aspect to the cultural production of knowledge in the new anti-apartheid movement. Indeed, the poster designed by Brazilian cartoonist Latuff for Israeli Apartheid Week 2009 was banned in several Canadian universities. This led to its publication in numerous newspapers. Many young activists are attracted by the vibrancy of a movement that produces original art without compromising a political message.

In short, the conscious attempt to foreground the Palestinian narrative allows Palestinians to speak for themselves and to raise all aspects of the Palestinian identity (such as art, culture, and so forth) is a key enabling factor in the growth and development of the movement's knowledge. Without this, the movement tends to fall into patterns that unconsciously reproduce the dominant narrative.

Nonsectarian Movement Building

A very important movement-building lesson of the new anti-apartheid struggle is the core belief in nonsectarianism. By sectarianism we mean a multiplicity of different (usually small) groups with few substantive differences between them who compete against each other for political audience and memberships. The roots of sectarianism lie in fetishizing minor programmatic differences and organizational forms ahead of the interests of the movement as a whole. Often this sectarianism is brought into the Palestine solidarity movement from competing left organizations or other movement divisions.

Key to the success of Palestine solidarity work in Toronto has been the conscious embrace of nonsectarianism and the notion that we must "put the movement first." It has not been common practice on the left to put aside disagreements and work together in broad organizations for the best of movement building. Many left groups seem to fall into two categories: those that will not work with anyone else because they lack their line on every political question, or those that see coalition work only as recruitment grounds for themselves. Part of building a new anti-apartheid movement is to produce an entirely different model that can bring together different groups and individuals who may have disagreements about tactics and the specificity of analysis, but who see the overall goal of challenging Israeli apartheid as being above these differences.

This is not an easy process and needs to be constantly worked at. But it is clear that the deepening of the anti-apartheid movement will not come from a handful of people, but rather through a broad effort that allows groups and individuals to take their own initiative, utilize their own creative energies, and organize themselves. Once again the simplicity of the BDS call and its rootedness in the Palestinian narrative helps to provide this focus. We do not need to agree on every historical detail or strategic question (for example, the BDS movement has consciously not taken a position on the One or Two States debate).[11] Rather, we simply act in unity around the call from Palestine, attempt to work honestly with anyone who accepts that call regardless of past animosities that people may have had, and try to avoid the tedious sectarian divides that plague the left. We need to continue to develop the way we relate to each other and other groups, realizing that building unity in practice is our most powerful weapon.

This does not mean that we should not take positions and build our movements around a broader political vision that supports other progressive struggles. As noted earlier, this has been very important to strengthening the BDS movement. But radical politics should not mean isolationism and a refusal to work with those who may see the world differently. The art of politics is knowing how to work alongside people who do not agree with you on every point, and yet still retain the ability to show solidarity and articulate your politics in a clear fashion. Once again the left seems to fall into two camps: the supposedly "radical" who are incapable of building alliances and working with anyone who does not embrace their particular view of the world, or those who allow their politics to be swallowed up and muted by more liberal tendencies. Finding the right balance is difficult but essential. It also means emphasizing the importance of political education and drawing upon the radical traditions of the Palestinian struggle, from which there is much to learn and to be proud of.

What relationship does this have to knowledge production? The key lesson is that the production of knowledge is a collective process developed through a common political practice, not bequeathed to the movement from historical texts or pontificating from outside the struggle. Only through practice can we truly understand the world and our place in it. Sectarianism acts against this because it stifles the growth of a fully rounded movement and channels it into preconceived directions that are not concerned with the growth of the movement itself, but rather the petty competitive rivalries typical of small-group mentalities. We have to fully internalize the belief that we truly have something to learn from others and that by collectively engaging in building a movement we change ourselves and thereby our understanding of the world. A good dose of humility is essential to knowledge production!

Sustainable Knowledge Sharing and Learning

Part and parcel of being nonsectarian is creating an activist space that allows for creative knowledge production and allows people to participate with a sense of ownership and belonging to the space. Particular attention must be placed on knowledge and skill sharing to maintain an open, self-sustaining culture. The conscious attempt to pass on ways of thinking and ways of doing to new activists coming into the movement needs to be a central feature of our movement-building experience.

This vision is centered around seeing everyone within the movement as an important component with a potential contribution to make. There should be no division between "intellectual," "academic," or "activist"—rather, everyone should take responsibility for all aspects of the movement whether it be writing a leaflet, postering, or taking minutes. This runs up against the way we are socialized in capitalist society—particularly the division between mental and manual labor—and so needs to be approached consciously by the movement. It is easy to fall back into routines that reinforce a particular division of labor, and it is very common for this to take on a gendered aspect with men being the public face of the movement and women doing the behind-the-scenes organizing. Counteracting this tendency means consciously encouraging activists to take on all tasks, even those that they may initially feel uncomfortable doing. This does not happen by osmosis—it needs to be something that everyone attempts to take responsibility for and needs to be talked about openly.

An important part of this is to see skills not as special "gifts" that certain individuals have, but rather as learned skills to be passed on and developed through collective training. Not every member of the movement will necessarily like to do every task, but everyone should be able do everything. In this way we build a strong collective team that does not rely on just a few "stars."

One example of this in the Toronto context is the frequent use of "train the trainers" sessions, particularly in the labor sector. These are educational workshops for trade unionists so they can go out into their union locals and workplaces to do public education. This model has been very effective in building labor solidarity with Palestine and in creating an atmosphere within unions that encourages member education. It is an attempt to break the model of bringing in an "expert" or "academic" to lecture at workers, rather than have workers themselves responsible for learning an issue and being able to share the information in ways that relate to their own workplace.

Another key part of sharing knowledge is the pairing up of experienced organizers with less experienced ones who are newer to organizing. As the new anti-apartheid movement grows, people join who might have never organized before. Movement "buddies" can help answer questions, introduce new skills and lead people through the sometimes arcane world of organizing. It is absolutely necessary to keep the atmosphere inviting and unintimidating for new people. Unfortunately much of the North American left makes a habit of judging people who do not have the appropriate rhetoric, music tastes, dietary requirements, fashion sense, or lifestyle. This pushes many people away from organizing. Once again the guiding principle has to be "putting the movement first."

One way to help build a more welcoming atmosphere is to realize that knowledge is a collective responsibility, not the possession of a select few. The BDS movement in Toronto holds regular educational sessions at the beginning of meetings to discuss a variety of topics, ranging from questions about Palestine to Indigenous Peoples' struggles in North America, to Tamil liberation, and so forth. This attempt to provide a broader political education helps to build confidence in activists and sets them up for sticking around for the long haul. Without this broader political perspective it is very hard to remain active in any movement.

Open Structures for Participation

This fundamentally democratic and collective understanding of the ownership of knowledge and skills contradicts everything we are taught within a capitalist society. With this in mind, we will discuss the types of structures and organizing principles employed by Toronto-based movements in order to facilitate this knowledge-building approach within our activist work. There are two key interrelated issues here: open organizing spaces, and democratic and clear structures.

If we are to be successful in building a truly representative solidarity movement, then it must be able to attract people beyond a campus base. While students often form the core of many movements, without drawing in wider layers of the population, movements can ossify and become too connected to the ups and downs of student politics (not to mention the frequent turnover of the student body). In order to reach out beyond students, it is necessary to be multigenerational and to find ways for everyone to plug in and contribute regardless of their personal circumstances. One way we have tried to achieve this in Toronto has been through building sectoral (labor, campus, faculty, high school, cultural), task-based (e.g., research), and identity (e.g., queer) organizations. There is no magic blueprint, and these different committees and organizations tend to wax and wane, but the key idea is to find ways to involve people from a variety of backgrounds in spaces that they feel comfortable within and to which they can make a positive contribution.

The level of autonomy related to these organizations also poses the necessity of more centralized structures. There is nothing inherently undemocratic about structures and positions of responsibility within an organization. In fact, precisely the opposite is true. If structures are not made clear, then they inevitably tend to form around invisible hierarchies that usually reinforce class, race, and gender divisions. The challenge is to build a collective and accountable team that can make decisions and help to push things forward, while at the same time providing the spaces for people to take their own initiative, experiment with new ideas, and make their own mistakes.

Conclusion

Since the beginning of the Second Intifada in 2000, a new generation of Palestinian solidarity activists has mobilized in the streets, workplaces, campuses, and schools across Canada. Among the left and progressive movements, there is broad acceptance of the proposition that Israel is a "colonial-settler" state, to draw upon the title of Maxime Rodinson's (1973) classic work. This is an unprecedented achievement. Throughout the second half of the twentieth century, radical and progressive movements in North America generally refused to take an unequivocal stance in support of Palestinian liberation. Palestinian solidarity was marginal to the large mass struggles that took place in the latter half of the twentieth century, and the left commonly countenanced a supposedly "progressive Zionist" stance.

The new anti-apartheid movement has managed to break any veneer of a "progressive Zionism." As each day brings news of initiatives around the world to isolate the Israeli state through boycotts, divestment, and sanctions, Zionist propaganda is increasingly forced to respond to the campaigns of the Palestinian solidarity movement, rather than the other way around. This anti-apartheid movement did not come about overnight, but was grounded in new forms of knowledge developed by activists in struggle. The production of knowledge—both *by* activists and *for* activists—has been central to the movement's success and growth.

Our chapter has shown that this production of knowledge is something that has both a particular content (a systemic and radical view of the world, grounded in the Palestinian narrative) and form (collective, open, nonsectarian, and sustainable). The form and content are linked through the structures and practices of the movement. In other words, knowledge gained by movements is something that has to be consciously cultivated and built through the ways in which we organize ourselves.

As a final addendum, we would note that this analysis is naturally framed by a particular set of experiences and vantage point from which we have attempted to generalize. One of the major weaknesses of our movements, however, is the reluctance that many organizers show in thinking through and reflecting on these matters in a systematic and collective manner. We hope that this contribution will go some way toward encouraging that process.

Notes

1. The Apartheid Wall (Ar. *jidar al-fasl al-'unsuri*) is the name given by Palestinians to a network of concrete walls, fences, and militarized zones that Israel has been building in the West Bank since 2002. It surrounds Palestinian towns and villages, severely restricting movement in and out of these areas. The wall was declared contrary to international law by the International Court of Justice in 2004.
2. For a comprehensive analysis of Palestinian refugees, their origins, and current locations, see the material produced by Palestinian refugee advocacy organization, Badil, at http://www.badil.org.
3. A proper accounting of the Oslo process is now widespread. Two trenchant early critiques are Said (2000) and Usher (1995).
4. See the Boycott National Committee (http://www.bdsmovement.net) for the text of the Unified Call and regular updates of the BDS movement globally. Also see Stop the Wall Committee (www.stopthewall.org) and the Palestinian Academic and Cultural Boycott Initiative (www.pacbi.org).
5. During the 1970s, for example, Israel abstained from fourteen UN resolutions condemning South African apartheid. See Chris McGreal (2006) for further examples of the close relationship between Israel and South African apartheid.
6. We should also note here the contributions of South African anti-apartheid activists. An important turning point was the 2001 Durban Conference Against Racism, at which South African activists led mass demonstrations making the links between Israeli and South African apartheid.
7. See "Towards a Global Movement—A Framework for Today's Anti-Apartheid Activism," published by Stop the Wall, at http://stopthewall.org/activistresources/1496.shtml, and the material available at www.badil.org (particularly issue no. 38 of *Al Majdal* magazine, which contains an excellent overview of the apartheid analysis as well as a description of campaigns across the globe).
8. For example, securing a divestment motion at the union level requires sustained work to convince the membership of Israel's apartheid character. Sustained educational campaigns produced a new type of awareness of Israeli apartheid that pushed people to do more than provide charity, but rather responsibly participate in ending the apartheid situation.

9. These comments are not meant to denigrate the important contributions of many independent, nonsectarian leftists who have played an essential role in providing links with other struggles and a broader political analysis through which to build the anti-Apartheid movement.
10. As was common with all colonial ideology, the early Zionist movement often portrayed the existing population of Palestine as simply nonexistent or, alternatively, as "backward" and needing "improvement" by more advanced "Western civilization." In 1914, Israel's first president, Chaim Weizman (1983), said, "There is a country which happens to be called Palestine, a country without a people, and, on the other hand, there exists the Jewish people, and it has no country." The same theme remains a dominant feature of much Zionist ideology today.
11. Proponents of the Two-State solution argue that a separate Palestinian state should be established in the West Bank and Gaza Strip alongside an Israeli state. The One-State solution argues for a single, unitary state across the whole of what is today described as Israel, the West Bank, and Gaza Strip without regard to ethnic or religious differences. Within the latter, a range of visions exists. These include a bi-national state, some type of federal solution, or a democratic, secular state.

References

Colas, A. (2001). *International civil society: Social movements in world politics.* Cambridge, MA: Polity Press.
Keck, M. E., & K. Sikkink. (1998). *Activists beyond borders.* Ithaca, NY and London: Cornell University Press.
McGreal, C. (February 7, 2006). Brothers in arms—Israel's secret pact with Pretoria. *The Guardian,* p. 10.
Mihesuah, D. & and A. Wilson. (2004). *Indigenizing the academy: Transforming scholarship and empowering communities.* Lincoln, NE: University of Nebraska Press.
Mohanty, C. (2003). *Feminism without borders: Decolonizing theory, practicing solidarity.* Durham, NC and London: Duke University Press.
Rodinson, M. (1973). *Israel: A colonial-settler state?* New York: Monad Press.
Said, E. (2000). *The end of the "peace process": Oslo and after.* New York: Pantheon Books.
Smith, J., C. Chatfield, & R. Pangucco. (1997). *Transnational social movements and global politics: Solidarity beyond the state.* Syracuse: Syracuse University Press.
Usher, G. (1995). *Palestine in crisis: The struggle for peace and political independence after Oslo.* Transnational Institute: Pluto Press.
Weizmann, C. (1983). March 28, 1914. In *The letters and papers of Chaim Weizmann,* Vol. I, Series B, ed. B. Litvinoff, pp. 115–116. Jerusalem: Israel University Press.

CHAPTER 7

The Subjectivation of Marriage Migrants in Taiwan: The Insider's Perspectives

Hsiao-Chuan Hsia

The phenomenon of marriage migration in Taiwan began in the mid-1980s when it moved from the supposed periphery to the semi-periphery in the world system. Most marriage migrants decide to marry Taiwanese men because they hope to escape poverty and turbulence in their home countries, which are intensified by capitalist globalization. According to statistics released by Taiwan's Ministry of Interior, as of October 31, 2008, there were 411,315 foreign spouses (30.6 percent from Southeast Asia and 63.32 percent from mainland China) in the country. Ninety-two percent of these foreign spouses are women. Among the women from Southeast Asia, 64.1 percent are from Vietnam, 20.7 percent from Indonesia, 6.7 percent from Thailand, 5 percent from the Philippines, and 3.5 percent from Cambodia. Marriage migrants mainly marry farmers and working-class men (Hsia, 2004). Most arrive without knowing much Chinese or other languages commonly used in Taiwan, which leads to greater isolation. These women are constrained by stressed economic conditions, lack of social networks and support, and discriminatory practices in everyday lives, policies, and laws (Hsia, 2009).

To protect the rights and welfare of the immigrants and migrants, a group of organizations concerned about im/migrants' issues established the Alliance for Human Rights Legislation for Immigrants and Migrants (AHRLIM) on December 12, 2003. AHRLIM is Taiwan's first alliance campaigning specifically for immigrants' rights and welfare. Many NGOs handle welfare cases of marriage migrants but rarely stage contentious action against government policies and laws. AHRLIM is currently the only alliance that aims at changing immigration policies and laws. After years of AHRLIM's struggles, several significant changes have been achieved, including the November 2007 amendments of the Immigration Act and the Statute Governing the Relations between the People of the Taiwan Area and the Mainland China Area on June 9, 2009, the two most crucial laws affecting marriage migrants in Taiwan.

In every AHRLIM campaign, marriage migrants play significant roles. Their presence at various protest actions not only catches public attention but also establishes the legitimacy of the AHRLIM-spearheaded immigrant movement. On September 9, 2007, hundreds of marriage migrants from Southeast Asia and mainland China joined hands in an AHRLIM-organized protest rally in Taipei against the financial requirement for naturalization[1] imposed by Taiwan's government. This rally assembled in front of the Executive Yuan (the executive body of the central government in Taiwan), marched to the Presidential Office Building, and ended with another picket in front of the National Immigration Agency. This protest action was historic because it was the first time that marriage migrants from all over Taiwan took to the streets to oppose policies violating their rights.

While it is crucial for marriage migrants to be active in the advancement of the immigrant movement, most have had no experience in political activism or even simple participation in civil society associations. Moreover, unlike in North America, Europe, and Australia, where immigrant networks are well established, marriage migrants in Taiwan, especially those from Southeast Asia, face severe isolation in a country where they do not even know the languages. Therefore, the issue of how these women can gradually transform themselves from being silenced in a strange and discriminatory host country to being vocal and active in the immigrant movement needs to be thoroughly examined. By examining my own long-term praxis, not only as an advocate, but also from directly participating in the empowerment of the marriage migrants and the making of an im/migrant movement in Taiwan since 1995, this chapter illustrates the learning process of marriage migrants, which in itself contributes to the knowledge production regarding the issues of subjectivation (Touraine, 1988).

Subjectivation and Societal Movement

Unlike dominant U.S. social movement theories, such as political opportunity, mobilization structures, and framing processes, Touraine pays more attention to issues of "subjectivation." To be distinguished from the commonly used term of social movement, Touraine (1988) introduced the concept of "societal movement," where historicity and subject are two key components. For him, historicity means how society acts upon itself to remake social relations and cultural models by which we represent ourselves and act. Societal movements are not merely groups of actors with specific grievances within institutions, but are marked by the degree to which they act upon the prevailing cultural model. Touraine is concerned about the struggle of social actors (the subject) over historicity, that is, who controls the terms of the cultural model upon which action is based. Moreover, the development of a societal movement is a process of transforming the "personal subject" into a "historical subject," which makes their mark on history by remaking the social relations and the cultural model that determine our identity. This process of subjectivation is the social action that challenges the existing social orders (Beckford, 1998).

This chapter echoes Touraine's concerns with "subject" and "subjectivation." Due to their social, economic, and cultural disadvantage, marriage migrants, referred to as "foreign brides" in mainstream societies, are often seen as victims and their agency is neglected (Hsia, 2008b). Some feminists challenge the prevailing assumption that foreign brides are simply victims or trafficked women (Constable, 2003, 2005). The focus has shifted to the agency of women whose cross-border marriages are seen as escapes from political, economic, and cultural constraints and courageous attempts to achieve a better future. However, we cannot ignore the structural limitations to individuals' resistance and should not narrowly perceive agency as merely individual escape from structural constraints. More important, these individuals may go further to form a "collective agency" that will transform these constraints. By doing so, they become "historical subjects." The individual escape may be seen as the expression of "personal subject," which by itself cannot transform historicity, that is, short of transforming into "historical subjects." This transformation will not occur automatically. Indeed, marriage migrants with multiple disadvantages in Taiwan often suppress their anger and resentment against unfair treatment or adopt the tactic of "exceptionalization" (Hsia, 2004); that is, to consider themselves as the exception from other "foreign brides" to resolve the inner conflicts created by the discrimination they face. This general status of "foreign brides" is what Marx (1978) called the class-in-self rather than class-for-itself.

The question of how the class-in-itself can be transformed into the class-for-itself has been much discussed and debated. Freire (1970) argues that through liberatory education, the oppressed are empowered to develop a critical consciousness of the world of oppression and commit to its transformation. Some critiques argue that Freire's theory is too linear and neglects the impacts of the unavoidable power relations between "teachers" and "students," and the dynamics among students of different social positioning (Ellsworth, 1989). Schapiro (1995) highlights the difficulty that practitioners of liberatory education face in overcoming entrenched power relationships especially when their students maintain an apparent passivity and look to them for leadership and guidance. The power relationships become more problematic when the teachers are from outside and have more privileges.

Moreover, several social movement theorists contend that the key to resolving differences among organizations with different social backgrounds is to clearly define "cause ownership" and the groups most directly affected, usually the underprivileged, should be acknowledged as having primacy and thus should take the lead in campaigns (Beamish & Luebbers, 2008; Stephen, 2008). Stephen (2008) specifically addresses the issues of collaboration between immigrant and nonimmigrant organizations in an alliance for immigrant rights; the ability of the nonimmigrant organization to follow the immigrant organization's lead is the key to develop trust necessary for successful alliance building.

In the United States and Western Europe, where immigration has a very long history and im/migrants have already developed solid networks and organizations, many im/migrant organizations have all the capacities to lead campaigns for their rights. However, Taiwan has a relatively short history of immigration. While im/migrant networks and political organizations are still very limited, the issues and difficulties facing im/migrants are very urgent under intensifying capitalist globalization. Therefore, Taiwan's movement organizations cannot simply ignore the conditions where immigrants' and migrants' human rights are egregiously violated and wait for them to build their own networks and political forces. However, issues of subjectivity for im/migrants are still crucial for creating a solid movement. Therefore, the challenge for organizations to build im/migrant movements is to simultaneously balance the need to tackle urgent issues and to empower im/migrants (Hsia, 2008a). Lacking strong social networks, and facility in the local languages, not to mention knowing how to access critical information such as laws and policies regulating their rights and welfare, there is a strong tendency for migrants and immigrants in Taiwan to look to the more privileged local people for leadership and guidance, and this dynamic needs to be addressed in the process of empowerment. In the following section, I will analyze my experience in the development of an immigrant movement in Taiwan to show how the marriage migrants are empowered and how the delicate issues of power relations between privileged local organizers and underprivileged marriage migrants are dealt with.

Demonstrating Marriage Migrants' Subjectivity in Immigrant Movement

The immigrant movement's legitimacy hinges on active participation of immigrants themselves. Many activists impose themselves as spokespeople on behalf of the marginalized, neglecting the subjectivity of the grassroots. Frequently it seems that those participating in a protest action are "mobilized" without fully knowing the issues at stake. It is therefore not surprising that Ms. Chen, a college-educated Vietnamese marriage migrant, wrote angrily to a newspaper questioning the legitimacy of AHRLIM's protest action against an anti-immigrant policy in July 2005, where marriage migrants from Kaohsiung (in southern Taiwan) joined marriage migrants from Taipei in a rally (Chen, 2005):

> We [marriage migrants] simply hope for good health, happiness, and being ourselves quietly. Especially for those foreign spouses whose lives are mostly confined to the boundary given by the husband's families, this is really the only simple and practical wish for them... But the newspapers reported that a group of so-called foreign brides took to the streets to protest against the Ministry of Education. These simple and innocent sisters probably could not have even entered the Ministry of Education in their home countries, and could not have even gone outside the front doors of the husband's family, how could they possibly travel from Kaohsiung to protest the Ministry of Education in Taipei?.... I experienced the worst pain and sorrow since I moved to Taiwan, not only because of the feeling of humiliation by the fact that foreign spouses are manipulated, but also because of my witnessing of the craziness of the Taiwanese... We foreign spouses are humans and we are not as despicable as to be treated as pigs, dogs or cows.

Being critical of such common practices of "speaking on behalf" of the mass in many social movements, AHRLIM has been conscious about the issues of subjectivity from the start. For eight years before AHRLIM was established, a grassroots organization, TransAsia Sisters Association, Taiwan (TASAT) had been empowering and organizing marriage migrant women. Consequently the subjectivity of marriage migrants had been gradually developed in the process, upon which the "legitimacy" of this immigrant movement is based. At AHRLIM's first protest, marriage migrants organized by TASAT were on the frontline. These women became significantly more active afterward, often participating in AHRLIM activities, speaking at protests or press conferences, and sharing their experiences and opinions.

TASAT's chairperson, Yudrung Chiu (2005), originally from Thailand, wrote a response to Ms. Chen, entitled "The Southeast Asian Sisters Already Rise Up":

> I used to have the same wish as Ms. Chen: simply hope for good health, happiness and being ourselves quietly. But now my thoughts are quite different. Aside from having good health, happiness and being myself quietly, I also want the rights I deserve.... One thing needs to be clarified to all is that we are not the "dogs, pigs or cows being manipulated by the Taiwanese" as Ms. Chen described in her article. We knew the issues very well before we went to the protest.... Our brains are clear!... And we did not just learn about it one or two days before the protest. In the past few years, we have been discussing various situations in our lives and our rights and welfares... and... how we can make the governments aware of our difficulties and our demands. We are willing to do things to demand for our own rights.... The sisters could stand up because we have the TransAsia Sisters Association in Taiwan.... By working together with Taiwanese sisters and Southeast Asian sisters, we hope to help more Southeast Asian sisters to stand up!

To demonstrate the subjectivity of marriage migrant women takes a long process of empowerment. While TASAT was formally established in December 2003, it originates from the Foreign Brides[2] Chinese Literacy Program, which I initiated at Meinung, Kaohsiung, in 1995. By providing a venue to learn Chinese collectively, this program helps women break their isolation and gradually build their subjectivities and collectivity. The program is not intended simply to assimilate immigrant women into mainstream Taiwanese society. It refuses to employ literacy as a tool to transmit an "ideology of accommodation" and reinforce a "culture of silence" (Freire, 1985). After much trial and error, the literacy programs gradually developed curriculums that combined the *Pedagogy of the Oppressed* (Freire, 1970) and the *Theater of the Oppressed* (Boal, 1979). Every class has a theme related to participants' needs (from bargaining in the markets to their welfare and rights), and includes icebreakers, discussions of issues, and learning keywords. In addition to improved Chinese abilities and self-confidence, the marriage migrants' networks, and those with other community organizations grew significantly (Hsia, 2006a). In 2002, in collaboration with volunteers, I formed another community base in Taipei. After eight years of grassroots-based empowerment of the marriage migrants and the local volunteers, we collectively established a national organization called TASAT.

In addition to empowering marriage migrants and Taiwanese volunteers, TASAT has tried to change public perceptions of immigrant women, through seminars, writings, paintings, theater, and documentary films. The women's voices can often help subvert the public image of them as submissive, problematic, and incompetent. In September 2005, the first collection of writings, paintings, and pictures of marriage migrant women, *Don't Call Me Foreign Bride* was published and attracted public attention. As editor, I noticed that a common response from readers was an appreciation of marriage migrant women's talents, and consequently their appreciation of multiculturalism and awareness of their own prejudices.

Subjectivation of Marriage Migrant Women

As much social movement studies scholarship suggests, potential for various collective activisms can be transformed into social movements only through efforts of formal or informal organizational networking (McAdam, McCarthy, & Zald, 1996; Tarrow, 1998). However, the isolation faced by marriage migrant women makes it very difficult to form informal social networks, much less formal organizations.

I now turn to analyze the principles and methods in the subjectivation process of marriage migrants employed by TASAT. I argue that in addition to Touraine's "personal subject" and "historical subject," a "communal subject" must be created in the subjectivation process.

Fulfilling Practical Needs to Initiate the Empowerment of Marriage Migrants

As Moser (1989) argues, the goal of gender-sensitive projects should be to address "strategic gender needs," which involve transforming oppressive structures, such as patriarchy. However, women in disadvantaged situations are often preoccupied by practical gender needs, such as childcare and making ends meet. To address only practical needs, however, is to reinforce oppressive structures (by perpetuating the ideology of "women's domain"). Moser maintains that effective projects must meet both practical and strategic gender needs.

By addressing marriage migrant women's practical need to learn Chinese, the literacy programs aim to create an opportunity for group dialogue by encouraging them to share their experiences, and gradually form their organization to speak up for their rights (Hsia, 2006a).

Learning language can fulfill practical needs while incorporating strategic needs. Freire (1970) argues that the first step of raising critical awareness (conscientization) is to break through the "culture of silence." For marriage migrants from Southeast Asia, knowing the local language is the precondition for breaking silence, and consequently forms the foundation of building an immigrant movement.

However, learning language does not automatically lead to fulfilling the strategic gender needs—to realize the roots of oppression. Many government and some NGO-run language programs aim merely to assimilate and domesticate migrant women. By emphasizing how to become a "good housewife" and "good mother," they neglect immigrant women's agency and the values of their cultures. As Freire (1985) observes, traditional literacy education only transmits an ideology of accommodation, reinforces the "culture of silence" that dominates most people, and thus can never be an instrument for transforming the real world. TASAT's view of literacy emphasizes the development of the learners' consciousness of their rights, along with an analysis of their position in the real world. We eventually developed a module to encourage women to share and reflect on their experiences, and to discuss issues and solutions by creative methods, such as sculpture and forum theater, to transform them from "Spectators" to "Spect-Actors" (Boal, 1992).

From "Personal Subject" to "Communal Subject"

With the help of our Chinese programs, marriage migrant women are no longer too shy to speak up and share in classes. However, despite common issues, they did not automatically develop a sense of "community" due to the barriers of differences in personalities, countries of origins, class, ethnicity, and educational levels, which often result in tension and conflicts. For instance, when women in our Taipei program were invited to contribute a presentation at one event, tensions and divisions developed in the process of preparation. Some complained to the local volunteers who served as "teachers," and threatened to leave the program if certain women continued to participate in the classes. The volunteers were very worried, and after discussion, we decided to bring the issue back for the immigrant women to face collectively. Using forum theater, a scenario regarding the obstacles facing the literacy program was played out. As problems arose, the play was stopped by the "joker" (which I played), who asked the audience to discuss how they would solve the problem and then implemented their plan of action.

To avoid the issue being "too close to discuss," the scenario played out was the potential division among the volunteers. After enthusiastic discussions, the immigrant women reaffirmed the group's importance, vowing to work collectively.

Our programs emphasize dialogue and encourage immigrant women to express their subjectivity. When encountering conflicts among individuals with different backgrounds, we still uphold this principle and facilitate a process whereby immigrant women can reflect on themselves, collectively come up with resolutions, and move further from personal to communal subjectivity. Without a sense of community, they cannot work collectively toward advancing their welfare and rights. Therefore, the "communal subject" needs to be added in the process of transformation from "personal subject" to "historical subject" in Touraine's theory of subjectivation.

From "Communal Subject" to "Historical Subject"

After several years of empowerment, the marriage migrant women from Meinung were no longer silent. They were willing to share their opinions in class. A sense of community, or transnational sisterhood among the immigrant women was gradually formed. However, the original goal of establishing a grassroots organization was still far from realization. Proposals of organizing collective action for their rights and welfare failed, because members were only interested in social activities, like outings or cooking. Some women lost their motivation to attend classes when they could speak fluently. The Meinung volunteers realized that the Chinese program itself was not enough for immigrant women to cultivate their collective strength because they still had to face daily problems at home after heated class discussions. The volunteers decided to organize a Hope Workshop in 2001, aiming at uniting the immigrant women through intensive discussions over two weekends.

The first half of the first weekend's workshop went very well. With dynamic methods, common problems were collectively summarized. However, when we requested that the women further discuss how to solve these problems, a strong sense of fatalism arose, with women saying, "It's all fate. We can't do anything about it!"

The volunteers were greatly frustrated and worried, knowing that the immigrant women could not rely on them forever and would have to face many difficulties by themselves. Finally, we decided to change the original plan for the next workshop and bring back the obstacles the Chinese program faced to the immigrant women, rather than relying solely on the volunteers for solutions.

We started the second workshop with a forum theater (see below for more on this technique), using the following scenario. Some volunteers enthusiastically phoned the immigrant women to remind them of the coming Chinese class, but everyone had reasons for absence and the volunteers were very frustrated. I served as the "joker" and asked every volunteer to express her feelings and worries. I then addressed the immigrant women saying, "It seems that all of the volunteers are very frustrated and tired, while the sisters (immigrant women) are no longer interested in participating in the programs. So it seems better to end the programs, so all will be happy." The immigrant women immediately objected. In response to questions posed by the joker, the marriage migrants collectively outlined the problems and discussed solutions. Eventually, a resolution emerged. Everyone decided to donate NT$300 (New Taiwan Dollars) to a fund and to find a meeting place where everyone could get together and organize activities. Because of this collective action, the immigrant women in Meinung found their own meeting space for the very first time. Their energy and strength could be combined and together they built a "new home," where they felt a sense of belonging.

After this, the Meinung women actively started many training programs, which inspired them to be independent and eager to help other underprivileged women. Organizers of the Chinese programs initiated a multilanguage hotline for marriage migrants with the Ministry of the Interior's Domestic Violence and Sexual Assault Prevention Committee. After over three years' discussion and negotiation, however, the committee transferred the project to a foundation, arguing that the organizers and immigrant women in Meinung were not professional social workers and may not understand procedures for handling hotlines. The volunteers were terribly sad, and after seeing the women's disappointment upon telling them the frustrating news, began sobbing and crying. At that moment, the immigrant women comforted the volunteers, and their strength helped calm the volunteers. Collectively, they all discussed and analyzed why they lost this project. Eventually, they realized that without a formal organization, it would be difficult to speak out collectively and make their demands heard by the government.

Through careful reflection and discussion, the frustration and anger over the hotline project became the turning point that helped the women understand the necessity of establishing a formal organization. They soon became actively involved in preparing TASAT's founding assembly, participating in every detail, from the mission statement to membership fees. Volunteers helped them draft TASAT's constitution, because the marriage migrants were still not familiar with the legal process of registering the organization at the Ministry of Interior. Later, they divided tasks into working groups.

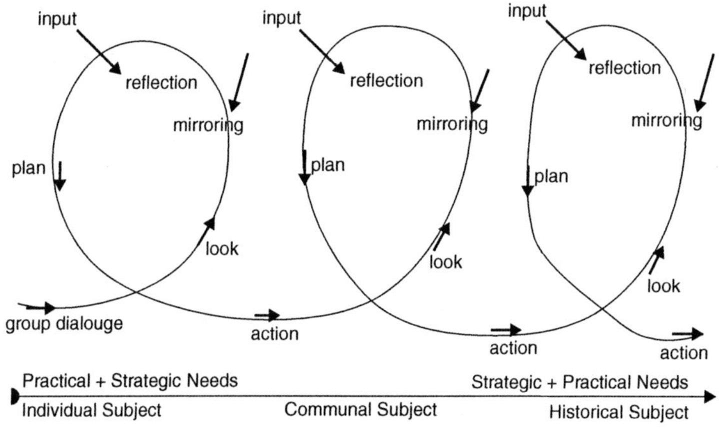

The Subjectivation of Marriage Migrants • 111

Figure 7.1 Subjectivation Process of Immigrant Women

TASAT is actively involved in AHRLIM to demand basic rights and push for changes in policies and public perceptions of marriage migrants. TASAT has also created a venue for marriage migrant women to join forces with other disadvantaged groups, such as migrant workers. By sharing its experiences, TASAT has broadened perspectives and expanded international networking. Through increasing networking with organizations from their home countries, marriage migrant women of TASAT have begun to understand capitalist globalization as a root cause of their escape from their home countries and to see the importance of transnational collaboration (Hsia, 2009).

Despite harsh structural constraints, the marriage migrants organized by TASAT have been greatly empowered and have significantly increased their participation in public issues and international networking. This process of subjectivation is summarized in Figure 7.1.

Keys to Transformation: Opportune "Mirrors"

The process of subjectivation is not linear and smooth. When encountering obstacles in actual praxis, methods to facilitate "breakthrough" and leap to a higher level of understanding and praxis are necessary. In *On Practice*, Mao (1937) articulated the dialectical relation between knowledge and practice. In the process of practice, at first a person sees only the phenomenal side, the separate aspects, the external relations of things, which is the "perceptual stage of cognition" (Mao, 1937, p. 262). As social practice continues, things that give rise to sensory perceptions and impressions are repeated many times; then a sudden change (leap) takes place in the brain in the process of cognition, and concepts are formed; they grasp the essence, the totality, and the internal relations of things, which is the stage of rational knowledge. Certain catalysts are needed for this leap to occur. Based on the experiences of empowering marriage migrants, I find that to help the women temporarily distance themselves and observe their experiences and practices more objectively is an important mechanism to propel transformation. Moreover, it is not to "teach" immigrant women how to understand things, but rather, to develop a "mirroring" effect by which they can see themselves, identify the problems, and find the solutions. The following are some important methods to create such "mirroring" effects.

Forum Theater

The techniques of the theater of the oppressed (Boal, 1979) are very effective in creating "mirrors," among which the forum theater is considered the most effective. Rather than providing an ending, the forum theater poses questions for the audience to discuss how to solve the problems.

Actual Practices

As Luria (1976) points out, experiences in collective action form the sociohistorical shaping of mental activity that significantly impact people's understanding of the world. Moreover, Foley (1999) argues that learning in social action is significant and empowering, largely informal, and often incidental.

For example, Peihsiang used to be one of the youngest, shyest, and most dependent members of TASAT. Through various trainings, she became more confident in sharing her own experiences in public forums. However, her perception remained relatively narrow and self-interested—she seldom took initiative, and mostly followed others. Peihsiang was tasked to speak at a forum about TASAT's history, principles, and methods of empowerment, because more experienced TASAT members were not available. Although she accepted the task, she was very anxious and relied solely on materials provided by TASAT staff. In helping her prepare, I briefed her about the audience and listened to her rehearse the presentation. Peihsiang said that she did not feel comfortable and confident because she did not personally participate in some parts of TASAT's history included in this presentation, and requested that I tell her what to do. Instead of telling her what to say in the presentation, I reminded her, "As the officer of TASAT, we need to learn broader perspectives, linking our own experiences to the mission of TASAT and the general immigrant movement. In the forum, you will represent TASAT, not just yourself, and the audience will learn and evaluate TASAT's work based on your presentation." Interestingly, Peihsiang suddenly appeared calm, and her presentation went very well and impressed the audience. Afterward, she told me that she was eager to know more about TASAT's history and all aspects of the im/migrant movement.

This vignette illustrates that we constantly encounter situations where the immigrant women look to us for answers. Instead of "teaching" them what to do, we strive to help them to situate themselves where they can link their personal lives to broader structural and historical contexts. From this realization, they can find ways to move forward.

Responsibility and Emotion: "Volleyball" Practices in Organizing

Previously, local volunteers accompanied marriage migrants in most situations so that they felt more secure in facing the general public. To overcome dependency on the volunteers, TASAT gradually gives responsibilities to marriage migrants. However, these responsibilities often lead to emotional reactions.

For instance, in preparing for TASAT's founding assembly, we divided the tasks for working groups, each with a migrant woman coordinator. Two weeks before the assembly, we met to discuss the list of recommended names for the coming election of officers. Instead of discussing who would be appropriate, the women complained how tired and useless they all felt. The volunteers, including myself, tried our best to energize and encourage them in the midst of depression and frustration. A voice of reflection suddenly came into my mind, "Why do I always have to shoulder the emotions of the marriage migrants? If I sincerely perceive them as my equals, why can't I share my emotions? If I continue to simply take on their emotions, would this stop them from growing? Is this a really equal partnership?"

These self-critical thoughts made me decide to break the usual pattern. I said, "All of us are tired. I think I am no less tired than any of you here. If we feel that it's not worth it to form the organization and all we get is tiredness, we can decide now to call off everything." Instead of suppressing my own emotions and feelings as I had usually done, I decided to reveal myself by walking out of the meeting after my comments. After I left, the marriage migrants were at first shocked because they had never seen me like that, and later started to share their bad feelings developed in the process of preparation, eventually deciding that it was crucial to form the organization, and made a list of recommended officers.

This experience helped me realize that although we had seen the importance for the marriage migrants to take up tasks and responsibilities, we failed to see that it is equally important for them to share emotions, especially frustration, that arise from taking responsibilities. As Foley (1999) suggests, learning in social action can transform power relations, but can also be contradictory and constraining. In the past, local volunteers took on responsibilities, while the marriage migrants perceived themselves as only "helping" the volunteers. Since the volunteers and organizers were conscious of their relative privilege, we unconsciously felt obligated to shoulder all the migrant women's emotions. However, this emotional dependence of the marriage migrants reflects a kind of "patronizing" relation, where they are protected by the local volunteers or organizers, just as children are protected by adults. This patronizing relation hinders the marriage migrants from transforming themselves into historical subjects, and reflects certain notions of heroism that we unconsciously hold. To ensure a truly equal partnership in TASAT, local volunteers and organizers need to learn how to share responsibilities and emotions with marriage migrants. This painstaking process involves the kind of interaction I call the "volleyball" practices in organizing: When marriage migrants toss up a ball of emotion to the organizers, we need to toss it back to them so that they have the opportunity to learn how to undertake responsibilities, and gradually become real team mates who can equally and collectively share responsibilities and tasks. Moreover, the local volunteer/activists need to learn that they also benefit from the advancement of the immigrant movement, because it develops a more open, multicultural, just, and democratic society for all. This realization helps break the mentality that the locals (us) are "helping" the im/migrants (them), and consequently develop a more genuine sense of comradeship between local volunteers/activists and im/migrants.[3]

Moving Toward the "Other"

Mainstream social movement theories argue that for the sake of effective mobilization, the framing process needs to distinguish "us" from "them"—that is, the boundary between "ally" and "enemy" (Tarrow, 1998). However, this is the politics of identity that Touraine (1988) criticizes for not being able to transform historicity.

As TASAT becomes more established, marriage migrants have developed a stronger identity with the organization. However, some members (locals and marriage migrants) fell into the trap of identity politics. For instance, some felt that other migrant workers should be blamed for the social stigma against marriage migrants because the former misbehave and come to Taiwan only to make money.

To transcend identity politics, TASAT has tried to help marriage migrants see the link between themselves and the so-called other through creating an environment where marriage migrants develop empathy for and build alliances with other disadvantaged groups. So, when interpreters are needed to help with migrant workers' cases, TASAT organizes marriage migrants to provide translation. TASAT organizers help explain the conditions of other migrant workers and the marriage migrants can develop empathy with them. Subsequently, TASAT members joined several rallies for migrant workers' issues. Moreover, TASAT joined the International Migrants Alliance (IMA), a global alliance of grassroots migrant organizations founded in 2008. By participating in IMA's activities, the marriage migrants expressed their appreciation of learning from and working in solidarity with grassroots migrant organizations in different countries, began to see themselves and TASAT as part of the broader global movement of im/migrants, and realized the importance of linking to the social movements in the home countries of im/migrants.

Conclusion

Employing Touraine's concept of "societal movement," and through a critical analysis of my direct involvement in the empowerment of marriage migrants, this chapter has particularly focused on "subjectivation" issues. Several lessons can be highlighted. First, subjectivation is a long and dialectic process, involving delicate interaction between emotion and reason, and sometimes even experiences retrogression. When encountering crises, through dialogue and reflection the subjects are gradually transformed to the next level. As Schapiro (1995) contends, some liberatory education theories assume what form of social action and participation is desirable, which gives the educators the ultimate ownership of truth and knowledge. Instead of assuming a predetermined path of development, TASAT perceives every bottleneck and crisis as great opportunities to collectively reflect and decide the next step. These bottlenecks and crises are crucial in the subjectivation process where the marriage migrants transform themselves to the next level of subjectivity.

Second, a mixture of disruptive and constructive tactics is needed to develop a mass movement whose primary movers are the disadvantaged mass. Some studies argue that "disruptive tactics" are necessary for social movement organizations. Gamson (1990) maintains that social movement organizations adopting "force and violence" are more successful than others.

Due to lack of resources, social movement organizations need certain disruptive tactics to attract public attention and push the state and society for changes. However, "constructive tactics" are also necessary to empower disadvantaged groups. Economically and socially marginalized, the disadvantaged mass can hardly afford frequent protest action. Therefore, the process of subjectivation needs to attend to their practical needs. Social movement organizations have the potential to frighten the disadvantaged if they always take up radical action, which will alienate themselves from the mass and weaken the movements. On the other hand, the historical subjectivity of the mass needs to be developed by certain forms of confrontations pushing for structural changes.

Third, scholar-activists like myself need to constantly reflect on our positions in social movements and transcend the tendency of reinforcing existing power relations. Touraine (1981) calls for the sociological intervention where sociologists should consider themselves as historical actors and representatives of real or potential actors greater than themselves, but at the same time should not be identified with the actors. For Touraine, maintaining distance helps delineate the social movement from the struggle and hence also to designate the social and cultural stakes of the conflict. However, this kind of distance may not be attainable in the social context where the gap between the haves and have-nots is severe, and consequently the intellectuals' access to critical information and their abilities of abstraction are needed to support the movements with more holistic and historical analyses. At the same time, we have to face the reality that intellectuals can be more easily co-opted by the status quo and may not have much at stake if they withdraw from the movements. Given these considerations, elsewhere (Hsia, 2006b) I argue that the intellectual should see him/herself as a "conscious wolf man," rather than a movement leader. The conscious wolf man is aware of his capacity to cause harm; therefore, before the full moon, he tries every means to avoid causing fatal damage. He constantly reminds people around that he might betray them and helps them learn all of his expertise and skills so that the people can carry on with their struggles after he eventually betrays. This metaphor of conscious wolf man illustrates that the role of intellectuals in the movement is not to lead, but to constantly empower more people to build their analytical capacities, avoiding the possibilities of inflicting a vital wound on the movement upon their future betrayal.

Notes

1. Marriage migrants were required to submit proof of financial security under strict guidelines, including a bank statement or official receipts for income tax

wherein the amount should be at least equal to 24 times that of the minimum wage. Many marriage migrants cannot apply for citizenship because of this.
2. "Foreign bride" is common parlance in Taiwan and reflects the discrimination against third-world women. Quotation marks are used to remind readers of the ideology behind the term.
3. The local volunteer/activists experience a similar learning and subjectivation process to the marriage migrants, which is beyond the scope of this chapter.

References

Beamish, T. & Luebbers, A. J. (2008). "Peace and justice: Learning from an alliance to stop a hot lab 'LULU' in Boston's South End." Paper presented at Annual Meeting of American Sociological Association, Boston, MA, July 31.

Beckford, J. A. (1998). Re-enchantment and demodernization: The recent writings of Alain Touraine. *European Journal of Social Theory* 1(2), 194–203.

Boal, A. (1979). *The theatre of the oppressed* [Trans. Adrian Jackson]. London: Pluto Press.

———. (1992). *Games for actors and non-actors*. [Trans. Adrian Jackson]. London and New York: Routledge.

Chen, H. F. (July 19, 2005). The arrogance behind the project of Taiwanese children by Mekong River. *Apple Daily* (Taipei), Public Forum page (in Chinese).

Chiu, Y. (August 1, 2005). Southeast Asian sisters already rise up. *Apple Daily* (Taipei), Public Forum page (in Chinese).

Constable, N. (2003). *Romance on a global stage: Pen pals, Virtual ethnography, and "mail order" marriages*. Berkeley: University of California Press.

———. (ed.). (2005). *Cross-border marriages: Gender and mobility in transnational Asia*. Philadelphia: University of Pennsylvania Press.

Ellsworth, E. (1989). Why Doesn't This Feel Empowering? Working Through The Repressive Myths of Critical Pedagogy. *Harvard Education Review* 59 (3), 297–324.

Foley, G. (1999). *Learning in social action: A contribution to understanding informal education*. London and New York: Zed Books.

Freire, P. (1970). *Pedagogy of the oppressed*. New York: Continuum.

———. (1985). *The politics of education: Culture, power and liberation*. [Trans. Donaldo Macedo]. Granby, MA: Bergin and Garvey.

Gamson, W. A. (1990). *Strategy of social protest*. Belmont, CA: Wadsworth.

Hsia, H. C. (2004). Internationalization of capital and the trade in Asian women: The case of "foreign brides" in Taiwan. In *Women and globalization*, ed. D. Aguilar & A. Lacsamana, pp. 181–229. Amherst, NY: Humanity Press.

———. (2006a). Empowering "foreign brides" and community through praxis-oriented research. *Societies Without Borders* 1, 89–108.

———. (2006b). The making of immigrants movement: Politics of differences, subjectivation and societal movement [in Chinese]. *Taiwan: A Radical Quarterly in Social Studies* 61(1), 1–71.

Hsia, H. C. (2008a). The development of immigrant movement in Taiwan: The case of alliance of human rights legislation for immigrants and migrants. *Development and Society* 37(2), 187–217.

———. (2008b). Beyond victimization: The empowerment of "foreign brides" in resisting capitalist globalization. *China Journal of Social Work* 1(2), 130–148.

———. (2009). Foreign brides, multiple citizenship and immigrant movement in Taiwan. *Asia and the Pacific Migration Journal* 18(1), 17–46.

Luria, A. R. (1976). *Cognitive development: Its cultural and social foundations.* Cambridge, MA: Harvard University Press.

Mao, Z. (1937). On practice: On the relation between knowledge and practice, between knowing and doing. In *The Selected Writings of Mao Zedong*, 1970, vol. 1 (pp. 259–273). Beijing: People's Publishing House. (In Chinese)

Marx, Karl. (1978[1847]). The coming upheaval. In *The Marx-Engels reader*, ed. R. C. Tucker, pp. 218–219. New York: W.W. Norton.

McAdam, D., J. D. McCarthy, & M. N. Zald (eds.) (1996). *Comparative perspectives on social movements: Political opportunities, mobilizing structures and cultural framings.* Cambridge: Cambridge University Press.

Moser, C. O.N. (1989). Gender planning in the third world: Meeting practical and strategic needs. *World Development*, 17, (11), 1799–1825.

Schapiro, R. M. (1995). Liberatory pedagogy and the development paradox. *Convergence*, 28 (2), 28–48.

Stephen, L. (2008). Building alliances: Defending immigrant rights in rural Oregon. *Journal of Rural Studies* 24(2), 197–208.

Tarrow, S. (1998). *Power in movement: Social movements and contentious politics* (2nd ed). Cambridge: Cambridge University Press.

Touraine, A. (1981). *The voice and the eye: An analysis of social movements.* New York: Cambridge University Press.

———. (1988). *Return of the actor.* Minneapolis: University of Minnesota Press.

PART II

Making Knowledge and Learning from Unions, Worker Alliances, and Left Party-Political Activism

CHAPTER 8

Learning to Win: Exploring Knowledge and Strategy Development in Anti-Privatization Struggles in Colombia

Mario Novelli

On Christmas Day 2001, several hundred workers began an occupation of the headquarters of EMCALI, the public service provider of water, electricity, and telecommunications in Cali, Colombia's second largest city. The occupation was a response to a government announcement to privatize the company. Thirty-six days later, the workers emerged victorious after signing an agreement to keep EMCALI public. During the occupation, marches, strikes, protests, and blockades were carried out in Cali by EMCALI workers affiliated with the trade union SINTRAEMCALI, with support from local trade unionists from the region and supporters from the local communities. In Colombia's capital, Bogotá, toward the end of the dispute, a second one-day occupation took over the headquarters of the government ministry responsible for implementing the privatization process. The second occupation was defended by local trade unionists from the region who surrounded the building. Meanwhile, in London, regular protests were held outside the Colombian embassy, and several live video link-ups took place between leaders of the British Trades Union Congress (TUC) and workers inside the Cali occupation, alongside a range of other transnational solidarity activities, with letters of support and solidarity arriving from all around the world.

The question of how this victory was achieved and disseminating that knowledge remains important, for it was done in the face of opposition from local elites, the national government, the World Bank, the International Monetary Fund (IMF), and several key transnational corporations (TNCs) with all of the structural, political, and military power that this entails. Its significance goes beyond Colombia because similar networks of actors have managed over recent decades to implement other processes of privatization across low-income countries, often in the face of widespread opposition, and generally to the detriment of the poorest of these societies.

This chapter argues that this victory did not emerge from thin air, but was a product of a long, hard process of learning and strategy development that, while firmly centered inside the union, extended out to local communities and other union and political activists, and transcended geographical boundaries to connect with fellow trade unionists, human rights activists, and solidarity movements in different parts of the planet. These networked links and alliances served, at particular strategic moments, to strengthen the militant activities of the SINTRAEMCALI workers, and in combination proved too much for a repressive Colombian state that was reeling under such relentless and multiscalar pressure.

First, I will raise some issues relating to the need to take seriously knowledge development among social and trade union movements. Second, I will develop an argument for the need to reconceptualize and expand our definition of trade union education through the notion of "strategic learning." Third, I will provide some brief background to the nature of trade union struggle in Colombia. I will then proceed to describe the different strategic learning processes led by SINTRAEMCALI and demonstrate both their interconnectedness and outcomes. Finally, I will reflect more broadly on the SINTRAEMCALI case and on insights it can offer for other movements.

Understanding Learning and Strategy Development in SINTRAEMCALI

Within global policy circles such as the IMF, the Organization for Economic Cooperation and Development (OECD), and the World Bank (2003), there is increasing emphasis on "knowledge," the "knowledge economy," and "policy learning," aimed at making capitalism more efficient, competitive, profitable, and dynamic. Oppositional forces such as trade unions and social movements often find themselves organizing against these objectives, substituting competitiveness and profitability with goals like social justice, equity, and equality. In this process of resistance and the development of alternatives, these movements are also engaged in counterhegemonic knowledge production, which plays a key role in determining success and failure. Yet research on the production, dissemination, and exchange of counterhegemonic knowledge among social movements remains marginalized (Novelli & Ferus-Comelo, 2010[1]).

Similarly, despite an important and growing literature on trade union renewal and resistance to neoliberalism (Herod, 1998, 2001; Lambert & Webster, 2001; Moody, 1997; Munck, 2002; von Holdt, 2002; Waterman, 1998), little intellectual inquiry has been devoted to exploring the microprocesses of how trade unions transform their activity and the role therein of education and learning processes. Part of the reason for this absence lies in a narrow conceptualization of education as taking place only in classrooms or through formal means. However, trade unionists, as political activists, do a great deal of learning through their activities and organizational and political work, and I argue that this praxis is an educational process in itself that sits alongside more "formal" learning spaces. Processes of "education" and "learning" are thus intimately tied to broader social processes, and can only be understood within that context. The question then is how best to understand the conditions under which unions and their members might change focus and direction and shift toward new patterns of organization and solidarity practices.

Popular Education in Latin America as Knowledge Resource

In searching for models of education that can be deployed to generate critical consciousness, Latin America provides a fertile ground. "Popular education," initially proposed by Paolo Freire, has captured the imagination of many social movements seeking to challenge elite structures of domination. In Nicaragua, the work of "popular educators" prepared the ground for the 1979 Sandinista revolution (Arnove, 1986), and one can see the legacy of Freire's work being continued throughout the continent.

Popular education is thus seen as one of the vehicles through which the process of challenging unequal structures can be achieved (Kane, 2001). It has at its center a fundamental commitment to social change in the interests of the "popular" classes. Furthermore, there is a direct relationship between this type of education and the institutions and organizations that have historically emerged to defend the interests of the poor, such as unions and social movements, and it seeks explicitly to strengthen these movements (Jara, 1989 cited in Kane, 2001). This organic relationship means that the "organization" becomes the "school" in which popular education takes place, and their "struggles and actions, their forms of organisation, their 'culture,' in the broadest sense, constitute the starting point of popular education and its on-going field of enquiry" (Kane, 2001, p. 13).

Thus "popular education" needs to be seen as not only involving formal educational events, but also as part of much bigger processes, which, though appearing "informal" and "arbitrary," is very deliberate. In this definition, both "popular education" events and the actual practice of "strategy development" and "protest actions" can be seen as examples of popular education, whereby the "school" (the social movement) learns. The first occurs whereby people consciously engage in educational practices (schooling), and the second whereby people are learning through social action. Foley (1999) suggests that a broad conception of education and learning should include formal education (taking place in educational institutions), incidental learning (taking place as we live, work, and engage in social action), informal education (where people teach and learn from each other in workplaces, families, communities, social movements) and nonformal education (structured systematic teaching and learning in a range of social settings).

To explore these processes we need to extend our gaze beyond formal training courses for activists and develop an analytical framework that is "open" and that allows for the rich diversity of ways that social movements (their activists and supporters) engage in learning.

Strategic Learning as Knowledge Production in Action

In order to do this we need to operationalize a notion of learning and education taking place within time and space through struggle and contestation. We need a dynamic model that can capture the dialectics of this process. Hay (1995) talks of "strategic learning" and offers a useful diagram to highlight it as a process in motion (figure 8.1).

Hay's model provides a way of theorizing the process whereby a trade union engages in strategic analysis, which in turn leads to strategic action, and then to intended and unintended consequences of action, and to further reflection/analysis and action. The agency of the union itself is immersed in structural constraints, which provide possibilities as well as limitations (Cox, 1996), while the strategic action often changes the structures upon which it acts. Throughout these entire processes, incidental, formal, informal, and nonformal education will be taking place and influencing strategy and outcomes to differing degrees. In the next section, I will begin the process of contextualizing the particularities of trade union struggle in Colombia in order to understand the conditions under which SINTRAEMCALI's struggle was waged.

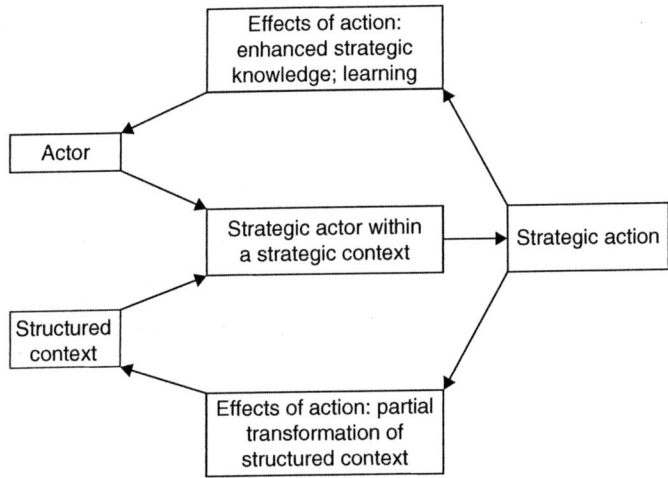

Figure 8.1 Strategic Learning
Source: Adapted from Hay, 1995, p. 202

Trade Union Struggle in Colombia

The SINTRAEMCALI conflict is firmly rooted in the context of neoliberal structural reforms that began in Colombia in the early 1990s. The shift to neoliberalism in Colombia led to wide-ranging transformations of social relations as the economy "opened up" to global markets. While Colombia has always been a highly stratified and unequal society, the neoliberal era intensified those inequalities. During successive governments throughout the 1990s, a range of policies based on a classic neoliberal recipe of decentralization, liberalization, privatization, and fiscal austerity were attempted, with varying degrees of success, as workers, *campesinos* [peasant farmers], Indigenous and black communities, social organizations, students, and trade unionists contested policy and engaged in defensive struggles to hold on to the limited gains won under the previous economic model (Ahumada, 1998, 2001; Sarmiento, 2001). For that resistance, many social movements paid a heavy price—Colombia is incredibly violent. State/social movement relations are permeated by violence.

According to government homicide statistics, during the 1990s Colombia had around 25,000 violent deaths a year (World Bank, 1999). While the vast majority of these can be attributed to "social" violence, according to Colombian human rights nongovernmental organizations (NGOs), there were between 5,000 and 8,000 "political" murders annually. While many of these deaths are linked to the decades-old civil war between Marxist-inspired guerrilla movements and Colombian state and "para-state" forces, the casualties go well beyond armed combatants and extend to broad sections of civil society and social movements. Within the context of what both the U.S. and Colombian governments term "a counterinsurgency war," the civilian population and nonarmed civil society groups have often been targeted, particularly by far-right paramilitary organizations ("death squads") linked to the Colombian state (Human Rights Watch, 1996, 2000, 2001). Union and social movement activity and the defense of economic, social, and cultural rights is generally interpreted as merely a cover for guerrilla activity, and thus leaders and activists are constructed as legitimate military targets (Stokes, 2004). This "dirty war" has led to more than 3 million people becoming internally displaced out of a population of 45 million. Since 1986, more than 3,500 Colombian union leaders and activists have been murdered (CUT, 2008) by paramilitary organizations, making Colombia the most dangerous place in the world to be a trade unionist.[2] SINTRAEMCALI was no exception, having lost seventeen members to political violence since 1997.

SINTRAEMCALI: Learning to Win

The push to privatize EMCALI began in the mid-1990s. Prior to that, according to Alexander Lopez, the ex-president of the trade union and architect of its renewal, SINTRAEMCALI "was just a calm, normal union working on normal grievances" (Podur, 2003). However the onset of neoliberal restructuring in Colombia produced a new environment that forced the organization to rethink its strategic direction. In response, a new and more militant group of workers began to take the initiative. The new leadership emerged from a rank-and-file committee that was born in the sewerage section of the company. After initially gaining success in improving working conditions for the sewerage workers, this committee began to push for greater influence in the union as a whole. In 1994, Lopez was elected to the executive committee, and within three years the balance of power within the union was won over to a position of radical opposition to privatization. From that point on the question became one of strategy. How could the union take on a highly repressive state and its transnational allies in Colombia's complex and violent conditions? Four linked strategic directions emerged that relied on the construction and strengthening of particular learning spaces where common understandings were built and new strategies and activities developed.

Worker/Community Alliance

One crucial strategy that emerged was the necessity of reframing the conflict over public services away from a trade union/state problem to one between the Cali community and the state. This necessitated a broad strategic alliance with local poor communities, but building this was not an easy task. There was a problem within EMCALI that could not just be explained by corrupt management practices. A certain level of complicity existed between workers and managers in the company's weak performance. This poor "service ethic" served to legitimate government and media accusations that the workers benefited from the good salary and benefits provided by the company at the community's expense. There was a need to change the attitude of workers and the attitude of the local community toward them. This required a sustained process of education on both sides, and a meeting point whereby workers and local communities could join together (interview with Alexander Lopez, 2002).[3]

This process of building up relations with the local community began in the mid-1990s, and aimed to raise the "political consciousness" of the local marginalized communities and to instil EMCALI workers with a sense of civic pride and responsibility. This work ranged from organizing public meetings, to workshops, to joining in local community demonstrations and protests and the setting up of a formal technical school based at the union's headquarters aimed at teenagers unable to afford secondary education. One central initiative was the setting up of *Minga Comunitarios* (an Indigenous phrase meaning "to come together"), whereby all EMCALI workers would give up one weekend each month and carry out repairs to damaged infrastructure in Cali's poorest areas. In order to cement this trade union/community alliance, new forms of democratic representation emerged whereby the union and community activists could develop joint proposals on strategy for the defense of public services. These initiatives served to develop strong bonds between workers and local communities.

Alexander Lopez talks of the obstacles that the union initially faced in constructing these alliances and the attitudinal transformation that later prevailed:

> In the beginning we had resistance from the community, because the community saw the workers as the problem, and we explained to the community who are the real problem, and where the neoliberal politics comes from, what is the World Bank, what is the International Monetary Fund, because the people don't know this. It has been a very dialectical strategy, very long, but also very beautiful and fruitful because...the

community takes time to make a decision, but when it makes it, it makes it forever. I believe that the people, yes, they have this logic, they don't recognize a lot of things, but they feel it when you speak, when you explain, and these type of lovers are not temporary loves but lovers that last forever (interview with Alexander Lopez, 2002).

The visible presence of the trade union in the neighborhoods served the purpose of "conscientization" in Freirean terms (Freire, 2000), where a "dialogical" process begins to emerge between organized labor and communities. As Lopez notes (interview, 2002), apart from attending to the real material needs of local communities and raising political consciousness, these were aimed at providing a "space" for challenging the attitudes of the communities and the workers toward each other.

Concretely, in the process of the development of the union/community alliance, the community brought with it a range of material and ideational resources that assisted the SINTRAEMCALI struggle. This network of organizations and individuals to differing degrees would show solidarity with the union's struggle, engage in marches, public demonstrations, provide material and political support to the union, apply pressure on local political representatives, and disseminate further an alternative reading of particular disputes from that depicted in the local press. On many occasions they have blocked roads, joined marches, attended public meetings, and shown solidarity with the union, not least in defending the perimeter of the series of occupations, and particularly the 1998 and 2001 occupations.

Plan PARE—Alternative Development Strategy

From the early 1990s SINTRAEMCALI had developed a collective understanding of the defense of public services as being a question of defending national sovereignty. However, what was lacking was a negative critique of neoliberal arguments on the inefficiency of state-owned public services, and a positive critique of how public services within EMCALI could be made more efficient. This led to the development of the "Plan PARE" (Plan for the Recuperation of EMCALI), a three-phase blueprint for the efficient management and delivery of public utilities developed by a combination of SINTRAEMCALI workers, sympathetic local technicians, lawyers, and financial experts. The first period, between 1995 and 1999, focused on investigating and exposing corruption within the company and highlighting that the crisis facing the company was due to widespread management corruption, political interference, and incompetence. This was carried out by groups of workers in each plant gathering data and evidence of what had happened. The second stage, between 1999 and 2001, resulted in the worker-led development of an alternative management plan for the company. Plan PARE was the product of workers' circles, initially organized at plant level, and later integrated across the company and assisted by technical experts brought in by the union who were sympathetic to the struggle unfolding. These laboratories of knowledge production allowed for workers' technical and tacit knowledge to be linked with professional and financial experts to produce a company plan based on economic efficiency, but also rooted in notions of public responsibility, equity, and social justice in regard to the key basic needs of water, electricity, and telecommunications. This plan was presented in late 1999 by SINTRAEMCALI, but was not accepted either by EMCALI management, the local council, or the Superintendent of Public Services. However, this was to change in 2001, when local elections led to a change in administration and a commitment by the new mayor to keep EMCALI in the public domain and to implement the plan. The third stage of the development of Plan PARE was its implementation, which began in May 2001. Due to the commitment of the workers and the quality of the plan, within eight months EMCALI had begun to show impressive results in cutting costs, improving efficiency, and reducing the company's debt. While SINTRAEMCALI celebrated the reversal of economic fortune, there was deep anxiety in government circles about the example that EMCALI might be setting. On December 24, 2001, the national government, drawing on extraordinary constitutional powers, sacked the new management team implementing Plan PARE and announced its intentions to push through with EMCALI's privatization (interview with Nelson Sanchez, 2002). SINTRAEMCALI's response was the occupation that began the following day.

Addressing the issue of the management of public services through the critique of corruption and the proactive Plan PARE allowed the union to move from defensive to offensive action. In this process the question of "who manages, who controls" was once again placed on the negotiating table (interview with Alexander Lopez, 2002). SINTRAEMCALI began to reconceptualize its role away from one of mediation to one of worker control of production. This shift in roles, while remaining partial, was highly emancipatory, as it began to throw up questions not only of workers' rights and benefits, but also of workers' control of production and the challenging of the labor/capital relation.

The development of Plan PARE also represented an intense laboratory whereby workers developed the new skills and knowledge necessary to manage the company and develop alternatives. Crucial to this process was the culture of learning that emerged within SINTRAEMCALI, and particularly the developing interest of workers in learning every intricate detail about the functioning of the company. The Plan PARE action research project meant that by 2001, the union had developed an enormous knowledge of the company's day-to-day running and financial difficulties, and was able to take its well-researched alternatives to a wide range of forums and debate alternatives.

During these interventions, networking and contacts would be developed through which joint political projects could be discussed and strengthened with independent political forces. Several sympathetic representatives of Congress and Senate would take up support of the case against privatization. Furthermore, at the municipal level, the union supported the mayoral campaign of Jon Maro Rodriguez, who upon election in 2000 began to implement Plan PARE. The mayor was able to do this because SINTRAEMCALI could deliver a credible and constructive alternative development plan for managing the company and a workforce that was prepared to make great sacrifices in order for the plan to succeed.

Mobilization Strategy

The third strategic trajectory of SINTRAEMCALI recognized the conflictual nature of any alternative development plan and the need to apply pressure on the government. The union recognized that the social forces intent on privatizing EMCALI would not be persuaded by words alone, and thus mass mobilization and direct action had to be deployed. SINTRAEMCALI's mobilization tactics included a combination of demonstrations, road blockades, and mass meetings that would often bring sections of the city to a standstill. However, from 1994 the union's trademark became the "occupation" of high-profile buildings, a highly symbolic tool that focused attention on an issue, raised the question of "control," but did not inflict material damage on the company or cause unnecessary inconvenience to the general public as the delivery of public services would always continue. The most spectacular of these were the occupations of the CAM Tower (the central headquarters of EMCALI) in 1998 for sixteen days and in 2001 for thirty-six days, actions that resulted in overturning decisions to privatize EMCALI.

While the success of mobilizations relied on local community support, they also depended on a solid core of members and supporters prepared to put themselves at great personal risk. As noted earlier, the internal transformation of the union was driven and sustained by a group of several hundred workers who emerged in and around the *Comite de Base* (rank-and-file committee). These members had a strong ideological commitment to the union's objectives and to a broader process of political change within the city and country. They were largely affiliates of SINTRAEMCALI, but also included several local unionists from other companies, community leaders, leftists, and radical human rights activists who saw the defense of public services as a key domain of struggle. From this group came the seventeen assassinated members, the forcibly displaced, and those personally affected by the policies of state and para-state repression. Many were actively involved in the construction of Plan PARE, and were involved in militant actions such as the occupations, blockades, and strikes since 1994.

When asked why members of this group were prepared to risk their lives for the defense of public services and thus for the union, Luis Hernandez, president of SINTRAEMCALI responded:

> I believe that you cannot buy this kind of loyalty, but you must construct it, and you construct it through a strong and responsible class leadership. I believe that those of us who have been leading this organization have demonstrated this and have built up a consciousness amongst the workers. This is a long process, but effective, very effective (interview with Luis Hernandez, 2002).

He highlights the role of the range of political education processes that many of these activists underwent:

> This loyalty has been developed from the training that our followers have received, and will continue to receive, because the good leader is in an evolutionary process, we are all in an evolutionary process and we are all learning, and developing and we will never have the final truth revealed, every day we need to learn more. That is why we are convinced that we need to generate more training and education. So that tomorrow it is not 500 workers but we can double that number or triple it, in the sense that we can keep generating more consciousness to struggle amongst the workers (interview with Luis Hernandez, 2002).

Many of this core group had friends assassinated since 1997, and had forged strong bonds with those comrades with whom they had engaged in direct actions and occupations. During these occupations, intense solidarities and bonds of trust were forged. This sense of collectivity and commitment provided the unpredictable element for the Colombian state, because these people had lost their fear, or at least had learned to manage it. This bonding and commitment, along with the new knowledge that developed during occupations, gave SINTRAEMCALI that vital edge in controlling those new spaces of protest.

SINTRAEMCALI and Human Rights and Solidarity Networks

The union's determination to militantly mobilize and occupy to defend EMCALI and its uncovering of high-level corruption placed it in direct confrontation with sections of the political and economic elite. In Colombia this meant facing the full wrath of state and para-state repression. The seventeen members assassinated, false imprisonment of workers under charges of terrorism, and a general and persistent barrage of death threats and assassination attempts led the union to adopt a range of preventative measures and work closely with national and international human rights organizations and networks.

In 1998, a prominent human rights activist, Berenice Celeyta, set up a human rights department inside the union to monitor human rights violations and to develop a strategy of human rights defense. At the local level, the human rights department trained local activists in human rights norms and mechanisms for self-defense. The training provided participants with methods of intervention into conflicts, how to set up a local human rights network among social movements, and a background on human rights legal frameworks at regional and global levels. These human rights training courses were expanded across the region in 2002 after securing international funding for the project which emerged as a result of the international solidarity networks. Graduates of these human rights courses and diplomas were able to act as intermediaries with state forces, government representatives and international bodies in specific conflicts and were aware of the functioning of this global infrastructure.

At the national and international level, the human rights department facilitated the involvement of major human rights organizations in SINTRAEMCALI's situation, and representatives of Amnesty International and Human Rights Watch (HRW) have intervened several times. The involvement of Amnesty and Human Rights Watch is understood by union leaders as providing a modicum of protection within which they carry out their activities and a means of addressing and revealing Colombia's "parallel state" of covert repression (interview with Berenice Celeyta, 2002; HRW, 1996, 2000, 2001).

Closely related to this network is the role of the supranational labor and human rights bodies, such as the International Labor Organization (ILO) and the Inter-American Court of Human Rights (IACHR). Assisted by national human rights organizations, SINTRAEMCALI took its case of systematic state and para-state persecution to the Inter-American Court of Human Rights. On June 21, 2000, the court found the Colombian government responsible for not providing adequate security for the trade union to carry out its legitimate duties. The court ordered the Colombian government to provide protection for the entire union leadership, including bodyguards, weapons, bullet-proof vehicles, and communications equipment (Inter-American Court of Human Rights, 2000). This protection, while not eliminating the danger, at least allowed some security and was a severe political defeat for the Colombian government who vigorously opposed the application of the measures.

More traditional international trade union activity was also developed alongside the Colombian *Central Unitaria de Trabajadores* (CUT), the biggest of the national union federations, and other national union organizations. With the backing of the British TUC, the CUT has also put forward motions for several years at the annual ILO conference in Geneva to call for a Special Commission of Inquiry into the human rights emergency for Colombian workers. The motion was finally accepted in 2006.

SINTRAEMCALI's human rights department has also been a fundamental catalyst in building transnational links with union, solidarity, and other activist organizations in the North. In the UK, strong international links began during Alexander Lopez's brief period of exile in London from October to December 2000. Since then there have been delegations, exchanges, and several invitations for SINTRAEMCALI representatives to talk at union conferences. SINTRAEMCALI forged strong links with the Colombia Solidarity Campaign, UK, which campaigns on human rights and political issues in Colombia, and in turn, several of its members have worked as volunteers in the union's human rights department. Since 2000, a continual process of communication and regular delegations has forged strong bonds between individual Colombian and British union and political activists. This strategy extended to Spain, Canada, the United States, and Australia, and included links with sympathetic politicians, human rights groups, NGOs, and social movements in each of these countries. The rapidly expanding Colombian exile community in these countries has strengthened transnational links, and many Colombians are founding members and activists in these solidarity organizations, as is the case in the UK.

Within the development of solidarity with SINTRAEMCALI, we can see the range of solidarity actions operating in both the more traditional forms and models of international trade union activity and the "new" labor internationalism, anticapitalist, and human rights domains. SINTRAEMCALI activists have participated in several World Social Forum events, European Social Forums, Latin American Social Forums, and have been invited to British trade union congresses of sister organizations and been present at annual ILO meetings. They also developed strong links with progressive NGOs, such as the UK's War on Want.

The union's human rights department provided a permanent point of contact for this international work, coordinating activities with international organizations, and producing and disseminating videos, reports, and urgent actions. The vast array of contacts and the experience that prominent human rights activist Berenice Celeyta has brought with her has ensured that the union capitalized on the potential interest of the international community in the situation of Colombia.

The human rights department provided the technical skills to improve the capacity of SINTRAEMCALI to challenge the Colombian state at a range of levels from local legal processes to the Inter-American Court of Human Rights. Central to this strategy was the need to create the necessary skills, knowledge, and materials to operate effectively in these new solidarity and human rights spaces. The human rights diplomas provided the skills for union and community activists. The production of videos, the detailed collection of human rights testimonies, and the support for delegations both to and from Colombia all reflected the union's commitment to educating both their own members in human rights and the international community of union and political activists about what was taking place in Cali and in Colombia more generally.

SINTRAEMCALI in Action

The thirty-six-day occupation demonstrates how these different strategies operated together to produce more than the sum of their parts. The trade union/community alliance mobilized on a massive scale on several occasions and provided a thirty-six-day cordon around the occupied building. The militancy of the rank and file meant that the several hundred workers inside the occupation stood firm and managed the massive psychological pressure, fear, and difficulties resulting from living in such a cramped and difficult situation. The knowledge and experience acquired from Plan PARE meant that the leadership had an authoritative argument to counter the government's critique of the company and the capacity to continue delivering public services throughout the occupation. Finally, the human rights and labor rights solidarity networks ensured that the occupation became an international labor issue, that international union and human rights organizations were centrally involved, and that the Colombian government was deterred from acting more violently for fear of international repercussions. In combination, on this occasion, this all proved to be too much for the Colombian state, which retreated to plan for the next confrontation.[4]

Conclusion

Central to SINTRAEMCALI's success was the long and hard work put into constructing these different strategic learning spaces. Formal, incidental, informal, and nonformal learning took place, and were firmly rooted in a political diagnosis of the particularities of the SINTRAEMCALI struggle in a violent and repressive Colombia and an increasingly globalized world. The strategy development and learning processes were dialectically linked to each other and developed out of the dynamics of the struggle. However, what was constant was a recognition of the importance of building alliances between those left out of the benefits of neoliberal globalization—locally, nationally, and globally—and the need to develop among its supporters the intellectual and political skills to challenge the neoliberal onslaught in a range of different ways. In doing so, this small union of 3,500 members provides us with a clear example that while over the past thirty years, neoliberal globalization has increasingly weakened the position of labor in relation to capital, this is not inevitable. Resistance can succeed, but it requires an openness to learning new things, thinking differently, building new alliances, constructing new alternatives, and creating a new kind of trade union subjectivity that, while continuing to be firmly rooted in local workplaces, is also linked into local community movements and alternative forms of globalization that can contribute resources (political, ideational, material) to the struggles ahead.

Notes

1. Novelli and Ferus-Comelo (2010) includes case studies of labor movement learning from around the world and develops a broad theoretical framework for exploring process of knowledge production in labor movements.
2. For an analysis of human rights violations against unionized educators during the same period, see Novelli (2009), *Colombia's Classroom Wars: Political violence against education sector trade unions*. Brussels: Education International.
3. All interviews for this research were conducted in Colombia between June and July 2002 as part of doctoral research on Colombia. See Novelli (2006).
4. See Novelli (2004) for a much deeper coverage of this whole issue; Novelli (2006) for a stronger political economy analysis located in research on social movement unionism; and Mathers and Novelli (2007) for a reflection on methodological issues related to researching radical social movements. The research was based on a four-year Economic and Social Research Council–funded study between 2000 and 2004.

References

Ahumada, C. (1998). *El modelo neoliberal.* [The neoliberal model] Bogota: El Ancora Editores.

Ahumada, C. (2001). *"Una década en reversa": Que esta pasando en Colombia?* [A decade in reverse: What is going on in Colombia?] Bogota: El Ancora Editores.

Arnove, R. F. (1986). *Education and revolution in Nicaragua.* New York & London: Praeger.

Cox, R. (1996). *Approaches to world order.* Cambridge: Cambridge University Press.

CUT. (2008). *Informe sobre los derechos humanos 2008.* Bogota: CUT.

Foley, G. (1999). *Learning in social action: A contribution to understanding informal education.* Bonn: IIZ-DVV; Leicester: NIACE; London: Zed.

Freire, P. (2000). *Pedagogy of the oppressed.* New York: Continuum.

Hay, C. (1995). Structure and agency. In *Theory and methods in political science*, ed. D. S. Marsh & G. Basingstoke, pp. 189–207. London: Macmillan.

Herod, A. (1998). *Organizing the landscape: Geographical perspectives on labor unionism.* Minneapolis & London: University of Minnesota Press.

———. (2001). *Labor geographies: Workers and landscapes of capitalism.* New York: Guilford Press.

Human Rights Watch [HRW] (Arms Project) (1996). *Colombia's killer networks: The military-paramilitary partnership and the United States.* New York: Human Rights Watch.

———. (2000 February). *The ties that bind: Colombia military-paramilitary links, 12*(1B). New York: Human Rights Watch.

———. (2001). *The "sixth division" military-paramilitary ties and U.S. policy in Colombia.* New York: Human Rights Watch.

Inter American Court of Human Rights (2000). *Medidas cauteleras 2000.* Retrieved from http://www.cidh.oas.org/medidas/2000.sp.htm

Kane, L. (2001). *Popular education and social change in Latin America.* London: Latin America Bureau.

Lambert, R. & E. Webster (2001). Southern unionism and the new labour internationalism. *Antipode 33*(3): 337–362.

Mathers, A. & M. Novelli. (2007). Researching resistance to neoliberal globalization: Engaged ethnography as solidarity and praxis. *Globalizations 4*(2), 229–250.

Moody, K. (1997). Towards an international social-movement unionism. *New Left Review* 1 (225), 52–72.

Munck, R. (2002). *Globalisation and labour: The new great transformation.* New York: Zed Books.

Novelli, M. (2004). *Trade unions, strategic pedagogy and globalisation: Learning from the anti-privatisation struggles of SINTRAEMCALI.* Unpublished doctoral dissertation, University of Bristol.

———. (2006). SINTRAEMCALI and Social Movement Unionism: A case-study of trade union resistance to neo-liberal globalisation in Colombia. In *Labour, the state, social movements and the challenge of neo-liberal globalization*, ed. S. Ludlam, A. Gamble, A. Taylor, & S. Wood, pp. 185–203. Manchester: Manchester University Press.

———. (2009). *Colombia's classroom wars: Political violence against education sector trade unions.* Brussels: Education International. Retrieved from http://download.ei-ie.org/Docs/WebDepot/EI_ColombiaStudy_eng_final_web.pdf

Novelli, M & A. Ferus-Comelo. (2010). *Globalisation, knowledge and labour.* London: Routledge.
Podur, J. (2003). Colombia's public services: An interview with Alex Lopez of SINTRAEMCALI by Justin Podur. *En Camino.* Retrieved from http://www.en-camino.org/node/1.
Sarmiento, E. (2001 April). El Neoliberalismo: Nefasto Experimento para Colombia. *Nueva Gaceta* (Bogotá), 88–97.
Stokes, D. (2004) *Terrorising Colombia: America's other war.* London: Zed Books.
von Holdt, K. (2002). Social movement unionism: The case of South Africa. *Work Employment and Society* 16(2): 283–304.
Waterman, P. (1998). *Globalization, social movements and the new internationalisms.* London: Mansell.
World Bank. (1999). *Violence in Colombia.* Washington, DC: World Bank.
———. (2003). Lifelong learning in the global knowledge economy: Challenges for developing countries: A World Bank report. *Directions in development.* Washington, DC: World Bank.

CHAPTER 9

Worker Education and Social Movement Knowledge Production: Practical Tensions and Lessons

David Bleakney and Michael Morrill

Introduction

This chapter explores some tensions underlying and restricting social movement learning and knowledge production within trade union and worker contexts in the United States and Canada. First, we ask how trade unions reflect and participate in the education models of their oppressors. Second, we explore some of the methods and practices that have evolved, been borrowed, or developed that aim to reach beyond repressive learning treadmills and models, and the ineffective learning experiences that almost every worker will tell you about if asked.

Our shared experiences as facilitators and organizers in working-class struggles situate us well in terms of educational practice with adult working people. In thinking through these tensions, and in our own practice, we borrow from social movements of the global South, Indigenous Peoples' and women's circles, workers, migrants, and freedom struggles from the past and present.

Although we come from different sides of an international border, there is a long and shared history of struggle within unions in the United States and Canada. Some workers share the same employers or international trade union leadership. Our communities face many of the same problems situated in consumer-driven Western societies on a finite planet under financial totalitarianism where the market is supposed to be the creator and fixer of all problems. We all live in an age of neoliberal capitalism.

One of us (Bleakney) has designed, written, and facilitated courses in trade union education since 1994. This included two years as a chief steward of a large mail processing plant in Toronto, one year as the local organization and education officer for the Toronto local of the Canadian Union of Postal Workers (CUPW), and for the past fourteen years as the elected union representative for education (Anglophone) for CUPW. The education program includes a national in-house four-week program, one-week and three-day seminars across the eight regions of Canada, and one-and-a-half day courses for more than two hundred locals.

The other (Morrill) has worked in U.S. labor and social change movements for four decades, including locals of the Service Employees International Union (SEIU) and the American Federation of State, County, and Municipal Employees. He has developed educational programs and curriculums to encourage both nonprofessional healthcare workers to gain power in their workplace and career ladder programs for low-income workers.

U.S. and Canadian unions have seen a steady hemorrhaging of membership, union density, and reduced bargaining power in the past two decades (Moody, 2007; Statistics Canada, 2007). "As recently as 1983, the U.S. union membership was above 20 percent, meaning it has dropped by some 40 percent in less than a quarter of a century" (Finlayson, 2007, p. 1). Due to reduced trade protections for domestic markets, overproduction, and offshore competition, workers find themselves with poor negotiating power and leverage. Such an environment can lead to hopelessness or perpetual self-victimization where much time is spent talking about deteriorating working conditions and what employers are doing. How much time is left to use crisis as opportunity or to stimulate grass roots and democratic organizing in collective and organic ways? Yet despite an especially hostile climate with downward pressure on wages and benefits, "unions remain stubbornly present" (Spencer, 2002, p. 11) and have the opportunity to contribute to grassroots education and organizing and connect with social movements.

What Is Union Education?

In the United States, academic and college programs provide the bulk of training for trade union members, although this is changing as many unions develop their own training and education funds. This is not the case for other forms of community and worker education at the grassroots. In Canada, most worker education is delivered by union leaders, staff, or worker-educators (workers booked off from their workplaces with lost wages paid for by the union to facilitate sessions with members of the same union).

If union education practices differ between the two countries, there are important similarities applicable to workers and communities on both sides of the border. Workers face an aggressive neoliberal climate and are on the defensive. Within unions great effort is spent on negotiation and consultation in the hopes of maintaining what benefits can be retained amid hypercapitalism and financial instability.

We concur with Newman (2002) that unions "by their very nature...are drawn towards the system. Their concerns are to do with wages and conditions of their members, that is, with economic considerations" (p. 165). Business unionism "is the sort of unionism that both treats the union as a business and emphasizes the need for the union (allegedly on behalf of the workers) and the employer to collaborate, at the expense of class struggle" (Fletcher & Gapasin, 2008, p. 251), while "in exchange for concessions from the employing class, unions commit themselves to maintaining a 'class truce,' hence limiting their objectives to operating within existing class relations" (Kuhling & Levant, 2006, p. 220).

This creates a tension—all unions are in a "business" relationship with their employers and face the conundrum of how much is spent on building the rank and file and the community on the one hand and how much on "defending" the worker on the other. This leaves space for employers to control or manipulate the agenda, choosing when and where to violate contracts and occupy union time in discussions, arbitration, phone calls, correspondence, e-mail, and negotiations. Union education processes focus on collective rights and how to defend them while developing a system of shop-floor advocates or "business agents."

Union bureaucracies must represent their members. But in Western capitalist societies like the United States or Canada, this can result in union leadership holding more meetings and dialogue with bosses than with the membership. Bosses can keep unions tied up and occupied. They frequently violate collective agreements. Every minute a union leader is in dialogue with their employer is time not spent with members or used to organize and agitate for critical changes and shifts in power. As a result, unions have come to "parallel the personnel departments of large corporations" (Taylor, 2001, p. 98). Unions are contradictory and negotiate terms of compliance.

There have been historical and current attempts to expand and move past traditional forms of education within unions and workers movements. But this is not without controversy and resistance. As Jeff Taylor (2001) notes, "In 1929 the American Federation of Labor (AFL) exerted its control over the WEB (Workers' Education Bureau of America) to ensure that trade union educational endeavors supported collective bargaining rather than attempts to change society" (p. 5). Worker education in the United States "underwent a shift in the 1930s and 1940s, from an older 'workers education' that stressed education for social change," which "was opened to unorganized and organized workers and was organized primarily outside of trade unions, to a newer 'labor education' that focused on training trade unionists to participate in collective bargaining" (Taylor, 2001, pp. 47–48). Around the same time, labor radio broadcasters on the Canadian Broadcasting Corporation (CBC) Radio became "integrated into the government's propaganda machine and began promoting labour-management cooperation" (Taylor, 2001, p. 37). In the United States, political pressure imposed through the Taft-Hartley Act required trade unionists to sign affidavits claiming "that they did not support the communist party" (Fletcher & Gapasin, 2008, p. 28). In 1949–50, the Council of Industrial Organizations (CIO) "expelled eleven...unions for refusing to sign the anticommunist affidavit" (Fletcher & Gapasin, 2008, p. 28). Walter Reuther of the United Auto Workers aligned himself with a "rabidly anticommunist" alliance to consolidate power of his faction, which "paved the way for the CIO's surrender on the anticommunist affidavit issue" (Fletcher & Gapasin, 2008, p. 28). Reuther also removed "alleged communists" and "sympathizers from paid positions within the union" (Taylor, 2001, p. 66).

In central labor bodies there remains a significant effort to channel constituent anger to a political party (for most unions this means support for the Democrats in the United States and the New Democratic Party in Canada). Perhaps it is a case of political pragmatism, but this comes at the price of disenfranchisement and conformity. Workers could make relatively pragmatic choices that will be in their interest, but we believe that we silence or obscure struggle by making this the sole avenue of worker rage. Sometimes this political bias and faith in the electoral process has been used by union leaders to silence workers. To cite one historical example, Taylor (2001) notes that "between 1948 and 1950 the Canadian Congress of Labour (CCL), under social democratic leadership, expelled the International Union of Mine, Mill, and Smelter Workers (Mine-Mill), the United Electrical, Radio and Machine Workers (UE), the International Woodworkers of America (IWA) and the International Leather and Fur Workers because their memberships continued to elect and support communist leaders" (p. 62). Further, "a non-sectarian, militant, autonomous and critical educational programme" was abandoned "in favour of narrow, union-based training that was under the control of union leaders" (Taylor, 2001, p. 68).

Working-class people today tell us that their experience in the public education system was mostly negative. In many cases it appears that it may have been psychologically damaging and left deep emotional scarring. Workers frequently report being traumatized by the isolation and streaming of the public education system. Undoubtedly there is strong support for public schools, but the question remains: In whose interest are they run? Our education models are based on scientism, quantification, and positivism. We are taught to be receivers of facts. There is little space for questioning or developing theoretical capacities and discussions about human relationships and freedom in a classroom.

This process has served the political and business class well. It has been adapted and fine-tuned to control and confuse working people for generations. From Friedrich Taylor and Henry Ford to present-day incarnations of "employee engagement" processes, there is no space for thoughts about workplace and community democracy, freedom and justice, or why things are the way they are and how we could change them.

The campaign to institute market and corporate thinking into public education and knowledge production is a stacked deck and by its very nature a failure for working people. The present model of what passes for "education" in a modern neoliberal society is a place where both Coca-Cola and the police have become frequent if not permanent residents. Researchers promote "instrumental rationality in education" and are "pursuing neo-liberal and human capital precepts that reduce teachers and students to objects being crushed under the burgeoning weight of neo-liberal globalization and its corresponding labour market demands" (Hyslop-Margison & Naseem, 2007, p. 2). In short, market-based educational approaches invite a culture of acceptance and compliance to a corporate world. The foundation of modern education pedagogy derives from a deliberately worker-hostile process used to support the perpetuation of an industrial age treadmill of functional workers under social control. The legacy of the "scientific" theories of philosophers like Comte and Spencer (Hyslop-Margison & Naseem, 2007) remain with us, and union education is not immune from this.

When trade unions rely solely on playing by the rules of this system and become uncritical participants in a legal process, they reinforce and acquiesce to this dominant paradigm. They want a piece of the pie without questioning the nature of it. Unions will rely on approaches that reinforce the quantification of knowledge through the transfer of facts. This denies and obscures the fact that both are subjective and relative to the vantage point of the viewer, as varied as human diversity itself. "Education either reinforces or challenges the existing social forces that aim to keep (workers) passive... either education domesticates people or it encourages their liberation" (Wallerstein & Auerbach, 2004, p. 7).

These tendencies play out in trade union structures that on the one hand challenge the authority of employees to do as they please, and on the other administer collective agreements, placing the union as both protector and player in a legal game generating huge costs and bureaucracies. The organization becomes a kind of buffer and receiver of angry member complaints, and little more than an extended workplace insurance policy. In this quantitative-oriented milieu, there is pressure for labor education to focus on a great many tools, courses, collective agreement administration, labor codes, health and safety instruction, and so on. But what about direct action, strategic planning, and the building of grassroots assemblies? Posing this question is a starting point. Unions are again stuck defending, on the one hand, hard-won rights that have generated new institutional frameworks and practices, and on the other, a desire for societal change. If this latter objective is to be taken seriously (and not as mere rhetoric), it requires a social movement where the actors play many roles in all their confident and creative diversity. This should surely be the aim of worker education in the twenty-first century. We must ask: What makes us reactive and docile? What can be learned of our past and how do we shift in new directions? How do we move beyond a place where such questions are viewed as subversive to a place of deeper learning, conceptual thinking, participation, and application?

There are enough unpleasant facts and human squalor that if information transfer was the only thing required (or the "banking" model, as Freire (1984) put it), we should already have reached a point of widespread rebellion. But in our experience, participants who become mere receivers of unpleasant facts can become disempowered or feel more hopeless about prospects for change than to begin with. Workers not permitted to be participants in critical reflection on their reality are not likely to be agents for change. Resistance is personal and social, and requires slipping outside the "quantitative box" into which we so easily, unconsciously, and compliantly fit. The process of using anger and action is valuable in worker-based trainings. However, when workers remain in the anger stage, worker-educators have done no favors to the transformative process. Indeed, resolutions to problems have been contracted out to union advocates, thus reinforcing the model of disempowerment and union paternalism, no matter how well intentioned. Facilitators must aim to assist participants in transforming rage into hope. With hope, action can follow. What is the purpose of worker education if not to assist people to act on their collective wisdom and experience? It gives people chances to be participants in profoundly democratic ways.

What Could Social Movement Worker Education Look Like?

How is union education delivered? Does it reinforce or deconstruct power relations? Union education is "about challenging the dominant power of employers and their supporters on the one hand, and building worker and union capacity on the other" (Taylor, 2001, p. 7). This causes a critical balancing act in negotiating and defending sectoral rights and becoming participants of societal change for the benefit of the many over the few. Deconstructing power is not just a question of economics or male-dominated hierarchies. In a time of the national security state, where perpetual manufactured fear consumes us, can there really be a "democratic" union environment if you are a Muslim man from the Middle East? What equality exists for a member of an Indigenous community living inside the United States or Canada, countries that both oppose the United Nations Declaration on the Rights of Indigenous Peoples? Education systems, our governments, and other cultural processes cultivate and maintain powerful and divisive myths underscoring the need for our work to reach far beyond the economic. Our practices around who gets heard and what is considered important inside a movement-building context reinforce and allow social fabrications and biases to play out. This topic deserves more attention and discussion than the scope of this chapter. Suffice it to say that we believe that practices inside unions often absorb and reproduce undemocratic practices and cultural biases.

Deconstructing the "banking" form of education for workers deconstructs power distribution in the union itself. The "teacher" is teacher no longer, nor a "leader" nor a "priest." Education "is not a matter of applying the formula and getting the answer ready-made. A well sprung-within, overflowing naturally into the outer world—this is education" (Vinoba, 1996, p. 10).

Effective worker education must facilitate a process that creates spaces for confidence building and action. Workers, as empty vessels of facts receiving more bad news, require something more than that which cannot so much be taught as experienced. Presentations of facts highlighting threats from financial markets or corporatism do not result in any measurable shift to the benefit of the worker or their communities. The reverse often occurs: The crisis becomes so overwhelming through the eye of the "individual" that it manifests as feelings of helplessness or comfort in "contracting out" a wishful hope for change to another party, like a union or political leader. Yet instead, the facilitator can adopt a new role as one who takes care of, provides support and observation for, and encourages and nurtures the group to rely on and develop their own capacity for analysis, reflection, and action.

Underlying this is our assertion that popular education is more than the practice of asking a few discussion questions, adding information, analyzing, and forming new analysis in a continuous spiral. How do people, specifically workers, learn? What is the way in which they learn? Worker-friendly education requires an inclusive process that does not subjugate the individual to dogma, and which encourages the participant to access ways of learning denied them in the formal education system. Fostering a sense of collective and personal responsibility deconstructs hierarchical and disempowering models of human interaction and practice.

Worker education, then, does not presume a straight transfer of facts and information from a central authority. People can and will act based on a deeper fulfilling personal and collective way when given the chance. It requires flexibility and qualitative methods. It is a place where facilitators become keen observers and listeners, intervening to maintain a learning space in innovative and challenging ways.

Worker education, applied in different ways, creates possibilities, including the casting of the tendency to situate ourselves as poor victims of circumstance. Victims repeat ineffective patterns of subservience and then complain about it where it is safe. This plays out clearly in conventions of central labor bodies, whether state or provincial. Convention resolution after resolution will have a slate of speakers eloquently complaining about what the "bosses" and "politicians" are doing to "us." The word "lobby" appears again and again in the texts of resolutions. No explanation is given as to what is meant by lobbying. It could be a safe place from which to pretend that action is being taken somewhere by someone. Is it a petition or a postcard campaign? Is it a "cap in hand" meeting with your oppressor over dinner? Or is it a request or a demand that will be won by rebellious struggle and collective action? Rarely is any collective action that is agitated by the grassroots able to break through a convention structure. They are spaces to blow off steam and complain.

To produce knowledge that is useful to a genuine social movement, workers should be able to prepare in a continuous culture of resistance that recognizes and improves on the little things we do each day in struggles within our different spheres of contact, whether they be with the "public," in the workplace, or among and with migrants, Indigenous Peoples, the political or business circles. The pursuit of narrow and relatively privileged sectoral worker interests is a recipe for self-imposed isolation that weakens the social movements and ultimately the unions themselves. The result of the narrow approach will be losing a key force in any broader social struggle: the power of organized labor.

Forms of communication processes and tools can be adapted to help nurture and sustain a continuous line of struggle, creativity, and power in the present. Activism takes many forms: All are important and not necessarily mutually exclusive. These roles can be played constructively in complementary ways or in the rote practices of learned and ingrained competition. Advocates, organizers, supporters, and a rebel spirit are all ingredients that can result in the building up of a movement with individual pieces working in tandem, or they become narrow vantage points of competitive self-interest resulting in gatekeeping or narrow frameworks of resistance that suppress differences in culture, language, race, and gender, and require conformity to formulaic displays of tolerance and words of "inclusion" so long as they are defined and managed by those in power.

In the context of union education practice to which we are committed, we must go beyond information transfer from a perceived higher authority to a place of creative planning, application, evaluation, recognition, and celebration of our victories. Parallel to this is a process of comfort zone expansion. Recognizing and naming this moment at the right time shows that we can cease to be victims at any time of our choosing and take whatever small steps each day to set our course. Through this we create and recognize new spaces for the collective voice of our communities and workplaces. Participants can be permitted, and even encouraged, to take these steps. Many already do and have not had the opportunity to name it or connect with others about it.

Radical anthropologist David Graeber (2007) writes:

> We must make our freedom by cutting holes in the fabric of this reality, by forging new realities which will, in turn, fashion us. Putting yourself in new situations constantly is the only way to ensure that you make your decisions unencumbered by the inertia of habit, custom, law, or prejudice—and it is up to you to create these situations. Freedom only exists in the moment of revolution. And those moments are not as rare as you think. Change, revolutionary change is going on constantly and everywhere—and everyone plays a part in it, consciously or not. (p. 9)

Consistent with Graeber, we turn now to illustrate how worker education processes can contribute to or embody social movement knowledge production, drawing on examples from both sides of the border.

1. SEIU Healthcare Pennsylvania, United States

If worker education is to have an impact on the social justice movement, it must be fully integrated into the culture of the unions from the local to the international level. It cannot wait until there is a crisis, or workers will not be adequately prepared. Education that explores power dynamics and how change takes place should be a regular part of organizing drives, contract campaigns and all aspects of the union's work. In the United States, the Service Employees International Union Healthcare Pennsylvania (SEIU HCPA) local has tried to do this, and the result has been workers who are more willing to engage in real struggle than most U.S. unions.

SEIU HCPA organizing drives have had a high success rate because they recognized the need to begin education around power and capital, starting with the first meeting. Organizers ask workers questions that they are able to answer themselves, based on their knowledge and experience. Questions such as, "Does the boss want you to have a union?" and "Why does the boss fight you?" elicit responses that help workers to recognize their existing well of knowledge. They do not need teachers or outsiders to lead them, but simply some guidance to effectively use what they already have. This method of organizing also inoculates the worker against the inevitable attacks by the boss during the next stages of the organizing drive, because the worker has begun to understand how collective action can change the power dynamics in the workplace. Just as important, this method of organizing establishes from the beginning a basis for taking direct action against the boss in the workplace and politicians in government.

Worker education continues in SEIU HCPA's annual meetings, contract campaigns, and formal training programs. In all of these forums there is the opportunity for education about social and economic justice in the workplace and in society.

In December 2008, workers at Aliquippa Medical Center in Aliquippa, Pennsylvania, were informed that the hospital was closing. The SEIU HCPA members at the hospital, including nurses and other healthcare workers, were then informed that they would not be paid for their last two weeks of work, despite the fact that executives and outside contractors would still be paid. Taking inspiration from workers at Republic Windows and Doors, who had taken over their Chicago workplace under similar conditions, SEIU HCPA members decided to occupy the hospital to demand their back pay. The hospital administration had the hospital workers removed by the police.

The workers were undeterred and continued a campaign of direct action while simultaneously pursuing their claims before the bankruptcy court. They organized demonstrations outside of the court hearings in Pittsburgh and traveled to Chicago to confront the owners of the hospital, Bridge Finance Group, where they were joined by hundreds of supporters.

This kind of direct confrontation stands in contrast to the typical response by many unions in the United States, which would be a staff-led series of negotiations and legal actions. The actions of the members of SEIU

HCPA were made possible by the education of workers that is integrated into all aspects of the union's activities.

2. CUPW at G8, Kananaskis, Alberta, Canada
During the 2002 G8 summit in the mountain resort of Kananaskis, Alberta, the Canadian government created a massive security zone and outlawed all nearby protest. Snipers, army, helicopters, and a security apparatus costing hundreds of millions of dollars prevented any activity from taking place in the vicinity of the resort. Legally sanctioned protest would instead take place nearly an hour away in Calgary. CUPW had booked a space nearby on First Nations land to train in direct action and challenge the summit perimeter. Two days prior to the training this arrangement was canceled due to intense government and police pressure. Subsequently, the training was held nearly two hours away and drew rank-and-file members from many communities across western Canada.

Participants had one week to prepare. Most had never participated in anything like this nor placed their bodies nonviolently on the line. There was fear. Most felt it was better to join the sanctioned protest in Calgary. A few (predominantly male) members said, "Let's go all the way," "Do what is necessary," and expressed the need to go to the G8 summit directly. Most reported having family and work responsibilities and feared arrest or losing their job. A decision was made to attend a rally in Calgary a week before the summit. Postal workers would go with pen and paper and ask the following of people in the street: "We represent workers at your public post office. The G8 leaders are meeting in a few days. Do you have any messages for the prime minister or anything you would like him to say to the other leaders? Your public post office will deliver these messages for free." Surprisingly there were no reports of anyone refusing. A public consultation process had taken place. The community was not being "told," but "asked," and workers would "serve" and "deliver." This built some links and understanding within the local community, many of whom had been conditioned to think that those protesting were violent hooligans bent on destruction. This refocused a message from the community back to the political power of state and police delivered by workers affected by neoliberal polices promoted by G8 leaders.

As the week went on there was a shift in the participants. Most had now decided to go to Kananaskis and confront the army and police in the name of democracy. Interestingly, many of the participants who had seemed more emboldened earlier in the week chose to form an affinity group and go to the legal rally in Calgary. In the end, the main road into the summit was paralyzed for the better part of a day. An aide to Canada's prime minister was forced to accept messages from the public in front of international media on a street barricaded by army and police. Postal workers carried their letters high and created a dilemma for the authorities. In Canada it is a criminal offence to delay the mail and all mail delivery to the prime minister is free. Therefore postal workers were only doing their duty under Canadian law. This created a paralysis in the state security forces, which waved the demonstrators through four checkpoints on a supposedly closed street. Why was this action important? It showed that working-class folks from small towns with no previous experience were capable of taking risks that they would not have taken previously. Those who participated experienced a feeling and life lesson that likely will not be forgotten. It involved the community; it was about serving the community and not preaching to it, and it showed that fear could be overcome with dignity and unity. In practical terms, it highlighted the costs of the summit versus the pittance pledged in aid to Africa while local farmers were experiencing one of the worst droughts on record. It brought together, like social movement education, the act and the community.

This outcome was the result of a more politicized rank-and-file-friendly union education program, and by learning through doing an action that involved a reclamation and use of collective power. Participants were able to work through their fear, emotions, and increased powers of observation and awareness, using cooperation and respect. Workers took the stage front and center. They were supported with spaces for creativity and dialogue to name and fulfill the objectives of the action. A more formal approach to learning would not have accomplished this transformation. We worked through anger and outrage in the training and used those emotional triggers to find stories, examples, and visions of hope before taking action.

There are many other examples. The cross-pollination of ideas and actions from social movements not specifically rooted in unions can be an important place of exchange, learning, and the building up of social knowledge production and constructing alliances across societies and borders. Recently there have been other examples with CUPW learning from other communities and thus creating a stronger bond of mutual solidarity. CUPW observed the traditional election process of the Barriere Lake Algonquins, who are struggling to maintain their sovereignty and learning. They taught a great deal about democracy, respect, and participation. Such examples indicate that the ground is fertile for alliances toward a different world.

Conclusion

None of our suggestions and experiences constitute a magic bullet. But they are signposts and directions that produce different results than the

hegemony of past practice of compliance or acceptance within union
cation. Unions by their very nature are engaged in class conflict wh......
they like it or not. Obscuring this fact and continuing to dampen down employer/worker tensions reduces the collective power of working people in relation to broader social movements and cultures. Reactive practices can be transcended by deconstructing the absorbed and self-destructive practices to a new engagement of communication and dialogue involving those actors themselves. Workers can do it. They just need a chance. They need the space, and unions are in a position to provide it when and where they choose to. Workers in the United States and Canada can ill afford to be victims or helpless, nor can they afford to isolate themselves from broader social struggles. Education has the capacity to bring critical thought into the learning process.

Working people aware of their own power and reaching beyond purely narrow interests are necessary ingredients to the social movement power needed to change course in a finite planet while building a different relationship to work and life.

Our collective history has been obscured but not erased. We know our stories and that real power exists in each one of us working collectively in alliances to build a better world. We have the invitation.

Worker Education Practices

Before we close this chapter, it somehow seems unfair to complain about what is missing in worker education and social movement production and not propose some modest processes that can lead to different outcomes. The following approaches can help set up a more effective learning experience and create spaces for social movement knowledge production. They are borrowed, nurtured, and developed from many places. The collective experience of workers does not have a copyright. None of these methods in and of themselves are a magic or simple answer. All contribute in some way to reaching outside the box and building confidence and action.

Ideally, the members of a union should be active participants who gain confidence and create change based on their needs and requirements. This occurs not only in the workplace, but also in the community and world at large.

Learning Channels

When workers in a group are asked to self-select how they learn most effectively, most will choose anything besides "hearing." Some learn best by doing, feeling, or seeing. Formal education functions primarily by "telling" others what they need to know and then giving some exercises to practice the "knowing." Effective social movement education requires techniques that touch on all learning channels. This provides a space for all, especially those rejected or isolated by their experience in school or workplace training.

The Circle

Use a circle for all kinds of reasons. In a real circle you see every face. You need each other when you stand up to injustice. A circle is a more honest and less hierarchical place. Mystics and aboriginal elders might say there is special power in a circle. Changing the chairs and functional rows into a circle changes the room dynamics and energy. Perhaps this is something our ancestors did but was obscured along the way by "scientism." It is also one that North American workers, conditioned to sit in rows and look for simple answers to complex questions, often resist at first. Providing a space for movement between a circle, open space, and tables for group work helps with diversity in learning channel techniques and group process.

Music

It is a good idea to have a music source to create ambience when people arrive, take breaks, and do group work. Music brings a new energy into the room. It can be a place where different cultures cross-pollinate each other, and participants can bring music that resonates with them. Drum circles can involve everyone.

Adapt

You may need to adapt or change discussion questions or tool applications to meet changing demands and needs of the group.

Bring Food

Participants will be well served by a chance to eat together as a group, and it keeps energy flowing. Valuable time is not lost seeking out the nearest fast food restaurant. Create a new space. This can be a good opportunity to support a women's collective or some other cause to do the catering. Having to leave for lunch and separate as a group is a wasted opportunity and poor use of limited time. Ensure in advance the importance of food being on time. Don't risk losing momentum or having lunch at the wrong time. It can really throw things off.

Stop Talking

Facilitators should become seamless and less obvious. You want to generate as much group ownership and work as possible. You may have lots of interesting anecdotes and stories, but limit them to where absolutely necessary and poignant. Use time for the benefit of the group. Creating new spaces involves letting a group breathe and come into their own.

Silence

Savor moments of silence. They are awkward if you let them be. Silence does not mean that learning is not happening. Some of our more lucid moments in thought take place in silence or when pondering something that we just heard. Embrace and nurture silence. Use it. Don't be a slave to time and quantitative approaches. Constant noise with no pause for reflection is another social component of neoliberalism and social control. Silence is another of our spaces if we wish to claim it.

Learning Contract

A class contract is not necessary in this kind of learning. A contract presumes that people do not know how to respect each other. By making a list of rote and predictable general rules, we make false assumptions that working people need to have their places of convergence socially policed. We respect each other all the time without making a lot of rules about it. Such rules limit us for a number of reasons. They become a place to act out, to tempt, which could be an expression of something else that is left unnamed. We reinforce the dated scientific model that denies our human capacity and obscures the applications of power that occur in every classroom of people and organization, consciously or unconsciously. We remove collective ownership of respectful process and contract it out to be policed by the "teacher," "enforcer," or "boss."

Building Allies

Another approach is one that presumes each of us knows how to respect the other or can quickly do so if given the opportunity. You can ask, "So, we know how to respect each other, right?" There is an ally (groups of two participants selected at random) who discusses the question in pairs: "What does respect look like? How do you respect me? How do I respect you? How do we respect a collective process?" This approach builds and expands upon individual reference points situated in a collective framework. Discussion in pairs can do more to build a container of respect than a list of imperial rules that some speak to on a flip chart and others use by rote. This approach of speaking about respect gives meaning. It also gives the group more tools and confidence up front to recognize and take things on should they arise.

The act of having allies check in with one another through the training allows for a confidential arrangement between two parties who have a chance to debrief and exchange learning goals or to discuss responses to elicitive and problem-posing questions. This process is one way to build a collective wall or container of safety around a group. A group that has this unspoken and unnamed "container" is ready to accept more conflict without paralysis and work through it. Participants may feel more confident to "speak up," because they know there is the support and safety to do so. Participants can learn a lot through conflict, but are often not given the chance.

Guests

For maximized learning, there should be no titles in the room. The mere presence of union leaders in the training causes other dynamics that may restrain the group somewhat. Another approach, if leaders must be there at all, is to have them commit as full and equal participants for the duration, or just have them speak and leave the group. If they choose to participate, that means full participation and not leaving the room constantly for other business or answering cell phones. It should be made clear that all are equals in this circle—no one takes more space than another.

Co-Facilitation/Observer Role

Co-facilitation is a powerful tool. The co-facilitator is not a passive bystander. He or she should apply active powers of observation to the process. What is the group doing? What are they thinking? What is being said? What is not said? Is everyone participating? Who is talking? Who is listening? Who is not? Who is on the margins? Who is in the mainstream? Does everyone have space to speak and participate? Is everyone being heard? Are there things your co-facilitator needs to know? The observant facilitator can add a great deal to the learning experience and can contribute in many unseen and unspoken ways that enhances the group and individual learning process.

Class Outline

Present the class outline in a general and visual way, using themes rather than specific times, exercises, and periods, which can become distracting for participants, especially for the class "timekeepers." Keep that part light and fun with a hint of mystery. Treat it like a journey and one that workers themselves have some control over its direction.

References

Finlayson, J. (2007). *Industrial relations bulletin* (Vol. 39, Issue 2). Vancouver: Business Council of British Columbia.

Fletcher, B. & F. Gapasin. (2008). *Solidarity divided: The crisis in organized labor and a new path toward social justice.* Berkeley and Los Angeles: University of California Press.

Freire, P. (1984). *Pedagogy of the oppressed.* New York: Continuum.

Graeber, D. (2007). Revolution in reverse (Or, on the conflict between political ontologies of violence and political ontologies of the imagination). Unpublished paper.

Hyslop-Margison, E. J. & M. A. Naseem. (2007). *Scientism and education: Empirical research as neo-liberal ideology.* New York: Springer.

Kuhling, C. & A. Levant. (2006). Political de-skilling/re-skilling: Flying squads and the crisis of working-class consciousness/self-organization. In *Sociology for changing the world: Social movements/social research*, ed. C. Frampton, G. Kinsman, A. K. Thompson, & K. Tilleczek, pp. 209–229. Halifax: Fernwood.

Moody, K. (2007). *U.S. labor in trouble and transition: The failure of reform from above, the promise of revival from below.* London: Verso.

Newman, M. (2002). Part of the system or part of civil society: Unions in Australia. In *Unions and learning in a global economy: International comparative perspectives*, ed. B. Spencer, pp. 160–168. Toronto: Thompson Educational.

Spencer, B. (2002). *Unions and learning in a global economy: International and comparative perspectives.* Toronto: Thompson Educational.

Statistics Canada (2007). *Labour force survey 2007* (1981–2006 data). Ottawa: Statistics Canada.

Taylor, J. (2001). *Union learning: Canadian labour education in the twentieth century.* Toronto: Thompson Educational.

Vinoba, B. (1996). Thoughts on education. Varanasi: Sarva Seva Sangh Prakashan.

Wallerstein, N. & E. Auerbach. (2004). *Problem-posing at work: Popular educator's guide.* Edmonton, AB: Grass Roots Press.

CHAPTER 10

Conversations on the M60: Knowledge Production through Collective Ethnographies

Biju Mathew

This chapter comes out of a persistent conflict I have encountered on the question of knowledge production and political work. I have been an organizer with the New York Taxi Workers Alliance (NYTWA) for more than twelve years now. The Alliance was born in early 1998 as nine organizers—Bhairavi Desai (who would become executive director), seven drivers, plus myself—sat over sacks of potatoes and onions in the basement of a Bangladeshi deli in SoHo and consolidated the gains of more than a year of organizing work and set into motion a new organization. Within months, on May 13, 1998, with fewer than 700 signed-up members (out of a total active workforce of more than 26,000 drivers in New York City), the Alliance launched the most successful strike action in the industry's history, with more than 98 percent of the cabs staying off the roads. Today the Alliance has more than 12,000 members, mostly yellow medallion taxicab drivers. The yellow medallion taxi is unique because its central organizational component—the medallion—is essentially a permit to put a taxi with the sole privilege of street hails on the streets of New York City (NYC). The current open-market value for a single medallion is more than $600,000. So, most drivers—in large part recent third-world immigrants—do not own these medallions, but merely lease (rent) the medallions (and taxis) from a close conglomerate of garages and brokerage houses.[1]

Ever since my book, *Taxi! Cabs and Capitalism in New York City* (Mathew, 2005), was published, I have consistently been asked whether I consider it to be an "ethnography." Sometimes I have run into academics who are surprised by the "amount of theory" in the book—as if the book's limits were somehow exceeded by the presence of theory. In a diametrically different reaction but along the same axis of theory, I have also been told that the book is a welcome addition to scholarship because of the ways in which it keeps both political action and social theory in play together. At the other end, I have been in many forums where I have been asked if the book was not biased and why it should be considered legitimate knowledge. In still other settings, I have been asked to develop and write more about a seemingly new invention—the category of "activist ethnography" (Chari & Donner, 2008). It is not difficult to dismiss some, if not all of these questions, as I often have. To the question about whether I consider the book to be an ethnography, I have often responded with a dismissive, "Why is that an important question? Doesn't it necessarily locate something called an ethnography as legitimate?" To be fair, I do understand that the question of ethnography/activist ethnography is being raised by friendly souls—trying to find ways of bestowing legitimacy on my work beyond what was already in place in the room. And yet it has rankled me. On the issue of bias, I have turned the question back rhetorically—"Do you consider the *New York Times* biased?" Better still, I have sometimes agreed wholeheartedly that the book is biased and asked the audience to read the *New York Times* alongside it for "balance."

Political Work and Knowledge Production

While many of the above reactions have come in university/academic spaces, and the subtext was always one of legitimacy of the knowledge produced in spaces of political work, my intent here is not to pick a straightforward and antagonistic bone with intellectual work conducted out of the academy. As a matter of fact, I would argue that far too often, activist spaces are dominated by a destructive anti-intellectualism that is very often advanced through an antitheory/antiacademy attitude.

So, while acknowledging that the university is an important space of knowledge production, I wish to premise this chapter on genealogies of intellectualism that acknowledge multiple spaces of origin and production of knowledge. Anderson (2007) argues that the idea of the intellectual and/or intelligentsia have, in the European context, three distinct spaces of production—the university, as in the German model; urban bohemia, as in the French *avant-garde*; and political work, as in the case, for instance, of the

Russian revolution. It is in this last category that I am most interested. How does one characterize and understand the knowledge production process from within spaces of political work? How does one think about knowledge production from within political spaces as different and distinct from other modes of knowledge production? To engage these questions, I want to focus on a two-year period, between 2005 and 2007, when the NYTWA was locked into a fierce battle with the City of New York and with segments of the industry ownership on the question of what was officially called the Taxi Technology Enhancement Program (TTEP), more popularly known as the GPS (global positioning system). The implementation of the GPS produced a complex and multilayered battle. My intent here is to subject the entire period to an intensive analysis to locate and analyze the nature of knowledge production during this struggle within the Taxi Alliance. Accordingly I start with a brief account of the context for the GPS/TTEP battle so as to make intelligible the debates and knowledge production issues that I explore later.

Tech in Taxis: A Brief Background

In April 2009, the Taxi and Limousine Commission (TLC), the mayoral agency that regulates the taxi industry in New York City, issued an RFI (request for information[2]) on a Phase 2 proposal for technology enhancement in NYC yellow cabs. This RFI came in the wake of what the TLC perceives as the successful completion of Phase 1 of the TTEP, in which it implemented a global positioning system (GPS) based technology package in New York yellow cabs. The Phase 1 package consisted of: (a) a GPS unit that automatically captured drivers' trip data; (b) a Driver Information Monitor (DIM) that allowed the TLC to text drivers; (c) a Passenger Information Monitor (PIM) that allowed vendors and vendor subcontractors to air TV programming and advertising aimed at the captive passenger in the backseat of a taxi; and (d) a credit card processor in the taxi's back cabin. The NYTWA responded to the RFI in June outlining its objections to the proposed Phase 2 plan, which included revisiting several claims of the Phase 1 implementation. The response to the RFI was strongly worded and was itself a reflection of the three-year period from 2005 to 2008 and the unresolved conflicts between drivers and the city/vendors on the question of this technology package.

Conversations on the M60

For most of its journey through Queens, the M60 bus stays along Astoria Boulevard and enters Manhattan via the Triborough Bridge and 125th Street in Harlem. It's a long ride from La Guardia airport (LGA) to Manhattan. Sometimes by the time we jumped off the bus at 125th Street and Lexington (to take the subway back to the office), it would have been a good hour. Often four of us rode the bus—Bhairavi, Javaid, Bill, and I—most often after a long stretch of outreach and mobilization work at the various taxi holding lots of LGA.

Early in August 2007, around a month before the NYTWA strike to stop the GPS, we boarded the bus as the light was beginning to fade. As the bus pulled out, one of us came up with the customary question, "So, how was it?" or, "So, what did you guys think?" We had each been in different lots for most of the day—Bhairavi and Bill in the Delta-US Airways lots and Javaid and me in the main lot. "It looks good," was Bill's opener. A few seconds of silence followed as each of us evaluated that feeling. Did we all feel the same? "Yes, I agree" Javaid said, "they all know the whole thing, and they are angry." Again the silence. Again the evaluation. The conversations always started slowly and then picked up momentum as the bus pulled out of the LGA airport complex. "It's exceptional," Bhairavi said, "the level of sophistication many drivers have on what this technology would mean." The analysis of the hundreds of conversations we had had through the day just rolled out—fast and precise. We had each been there for more than five or six hours. That meant that each of us had spoken to at least 200 drivers or driver groups throughout the day.

Bhairavi continued, summarizing several of the conversations with drivers:

> There is almost nobody who doesn't get it.... That what this GPS on the meter means is nothing less than handing over detailed information of drivers' earnings, their work patterns, their workday—data about themselves—over to the garages and brokers. Everybody knows that this technology will sooner rather than later become a new weapon and therefore more power to the owners.

Everybody nodded, for the larger implication was clear. In a situation where the owners already had more power, this technology would mean things would go further out of balance. Javaid jumped back in to say how many drivers were wondering whether they would be forced out of the industry once the owners and government use this data. Bill was getting emotional as usual. "Not only will they do all this, but they will make us pay for it too," Bill was shouting. We waited for him to calm down a bit. He continued:

> You know what is truly exceptional is that in spite of all the imbalances, in spite of all the inequities, drivers have continued to work because there

is this sense of independence...that they are their own boss, that nobody can check up on them, that they decide what is good and bad for themselves, and they can stop when they want to and start when they feel like...and now what is really getting clear to them is that with this technology they might lose all that...that the level of monitoring and therefore the level of control could begin to change.

Javaid nodded and said:

> There was one driver who said all this won't happen...that the Alliance and the drivers were being unnecessarily worried.... He said he had had a pilot project GPS in his car for months and nothing had changed.... You should have seen the reaction...everybody jumped at him.... Of course, they all said, they won't do anything now because they want you to accept the GPS...but once it is there, they can do what they want...maybe not today, maybe not even in six months, but definitely in a year, in two years...they can decide.... It was amazing.... Everybody just went after that one driver.... I had to get in the middle to stop the shouting match!

Another day, another M60 ride. This time it was just Javaid, Bhairavi, and me. Bill was holding the office together that day. Victor Salazar had also spent a large part of the afternoon at the main lot with me. But he had been doing this in the middle of his shift. He had just left, too, but with a passenger headed back to Manhattan.

"I had an exceptional conversation today," I started. "It was one of those conversations that leaves you completely dumbstruck." Everybody was all ears. "It was with this older *sardar* [Sikh] driver—not old, maybe late fifties, Surinder. He has been driving some ten years only." I paused to gather my thoughts and find an effective way to summarize Surinder. I knew it was a conversation that would never leave me. For one, I realized that the drivers had shown an amazing grasp of the technology. With only fragmented information about the actual system, and the most basic explanations about GPS technology, the drivers had woven together a most rich and comprehensive account of its capabilities. That the TLC claimed that most of these would not be implemented and that all this was just a "misinformation" campaign were beside the point. The drivers clearly had a better sense of what might come, based on a fine sense of what they saw as the technical capabilities of these systems and a realistic analysis of where they saw themselves within the power relations in the industry and the United States.

Bhairavi and I, intensely aware of the criticisms levied against the NYTWA, would constantly try to temper the analysis emerging from the drivers. "There is no proof that they will use it for collecting taxes," I had told a group of drivers the other day. It was one of the few occasions when the whole group erupted against me. "You can't wait till they do it. The question is, can they do it?" said Shikdar, a Bangladeshi driver. I had to admit they could. Javaid had watched me being cut down to size by the drivers! Laughing at me, he added, "Of course they can, brother."

Simply put, it was clear that little by little, with conversations that went from the airport lots to restaurants, from homes to mosques and gurdwaras, from conversations with friends and family in various parts of the world, to fragmentary conversations with passengers, the drivers had developed a detailed and full imaginary around the technology. Each time the NYTWA organizers were at the airport, intense discussions would break out in the lots. A large part of what happened, especially during the campaign's early phases, was that we, as organizers, would simply carry back a basket of new questions into the office. There, some of these questions would get answered through some new research, and some aspects of these answers would be part of the next flyer that the Alliance prepared. The simple truth was that while the Alliance had on occasion provided them with specific answers to a technology question, the drivers had led the Alliance in terms of building an understanding of this technology's full potential.

I began slowly:

> Surinder was saying that he does a lot to keep the brokers off his back. The brokers and the TLC, and now he has them not just sitting on his back, but also dipping into his pocket. He was so clear—that they were not just dipping into his pocket by charging the five percent on credit cards, but even more important, if he wants to get his own money, money he had earned by working all day, he has to depend on them and that's what he found completely unacceptable. He kept saying, "Not only do they have control over my credit card amounts, but you can see that in a couple of years, once the credit cards are in, seventy percent of our business will be credit. And then what happens? Each week, I have to go back to the broker and settle scores with him. He starts out from a situation where he had no power over my money to a situation where he now has almost complete control over my money." He kept repeating this, and he was angry. "Why should my money suddenly come under their control.... Why am I an independent contractor if they have control over my money and can make me bow before them for it."

The M60 was silent for a minute. Then Bhairavi said that she had an almost identical conversation with another driver in the Delta lot a couple of days before. "For the first time since leasing began, the owners have found a way to dip directly into the pockets of drivers," she said. The drivers are absolutely right."

The four of us used the bus ride regularly to reflect on our conversations in the lots. The other main organizers of the NYTWA, most of them Organizing Committee (OC)[3] members, would turn up at the office late at night to exchange notes, or do so more formally at the OC meetings held on the first Tuesday of every month.

That same night in early August 2007, Mamnun, Tipu, and Beresford came into the office, all with similar stories. The drivers were incensed about the five percent surcharge on the credit card. Others were concerned as to what would happen with the brokers and the garages. Tipu said that the 29th Street *masjid* (mosque) was a hive of activity as drivers discussed the fact that the TLC or anybody else who wanted to could figure out when each driver was stopping for prayers at which mosque. Harjinder called in to say that a lot of the owner-drivers were concerned as to what would happen to their taxes: Who, they asked, would issue a Form 1090? As the night wore on, the discussions began approaching a level of coherence. It was a conversation that had begun several months earlier in mid-2006. Early on, when any of us would be at the airport, the conversations were more sketchy, with many more questions than answers, many more silences than opinions. "Who were the vendors?" "Was any garage or broker getting a contract?" "Will we have to log in?" Who will keep the data?" "Will the TLC share the information?" and many more. Over a year these questions had transformed into a cogent imaginary around what the TLC's technology package might mean for the driver. Every round of conversations in the LGA and JFK airport lots came back via the M60 (or the A-train) into the office, and in the company of many other report-backs were consolidated through the nights. By the day, meetings with lawyers were unfolding to see how drivers' concerns over loss of control could be shaped into a lawsuit if needed. Each day's conversation led to the next day's outreach and mobilization, into new conversations with lawyers, special strategy sessions with the Central Labor Council (CLC), and then back into the lots, into restaurants and neighborhoods.

Beresford came into the office one night. "Professor," he said, addressing me, "tell me, can somebody mess with the software of this damn system and then mess with credit cards used in my car?" I was hesitant. "Why?" I asked. "Well, that's what a passenger just told me," he replied. "He was some computer guy from Wall Street."

Victor came in one day and offered his latest insight. "You don't need the GPS to run a credit card.... Is that true?" he asked. "Yes, the credit card system has nothing to do with the GPS," Bhairavi had said. "It's just that it's all being packaged together." "I know" said Victor, "a very interesting passenger told me.... He was from Canada and works for a bank." Clearly passengers were also part of the conversation.

In the end, the coherence of the critique as it had emerged from the mass base was clear. Somebody else—the TLC, the ownership, the vendors—was going to gain control over various aspects of their work lives. Through credit card monies, the owners would gain greater and greater control over how much a driver retained in his pocket. Data collected through the GPS tracking could potentially be misused by any state security agency. The entire taxation segment of drivers' incomes could be potentially formalized without any equivalent record of expenses, while the five percent service fee on the credit card allowed cab owners to double dip into the driver's pocket and thus break all leasing terms. These were all concerns that could be unified only under one head—control versus autonomy.

Flexible Subsumption as the New Condition of Labor

In the months to follow, and after each round of conversations at the airport lots and restaurants, at the office and in neighborhoods, the question of driver autonomy emerged as the single biggest concern.

Even as this clarity emerged, the GPS systems were fast entering the taxis. The Alliance had grown by leaps and bounds during the period of the mobilization. In some ways, it was precisely the exponential growth that had allowed us to hear the common concern around autonomy so clearly. Understanding this concern was crucial for the organization if it were to imagine its own work into the future.

Something very specific was happening in the taxi industry with the GPS system. Worker control had been an issue of great debate in the taxi industry ever since its inception. The ownership could never find an appropriate method of control over the drivers once they left the garage and were out of sight. The introduction of leasing in the 1970s, where the owner collected a rent up front and made, in formal terms, no further attempt to control the driver's work, was a pioneering departure from the accepted wisdom of labor control at the time. It preceded the general change in direction brought on by neoliberalism where an ever-expanding pool of contingent workers and independent contractors was being created. In other words, the taxi industry of the late 1970s was at the cutting edge of a paradigm shift in how labor was to be repositioned in the neoliberal economic formation.

The romance with independence, while objectively located in an exploitative relationship with the medallion ownership, was something that all of us had encountered since we began organizing in the taxi industry. In the mid-1990s, during the fledgling days of the NYTWA, when both Bhairavi and I were new to the industry, I remember several conversations that we had with each other, trying to get to the bottom of this idea of freedom that drivers seemed to cherish so much in spite of the hard and inhumane conditions of work. It was an ever-present category in the life of the Alliance—a kind of a steady refrain that was used to explain almost anything. And now, as the GPS system advanced into their midst, the question of autonomy had risen to the top, no longer a steady refrain but a vehement demand. It forced us at the Alliance to keep the question constantly in our discussion. The Alliance moved to file a law suit in the federal court in late September 2007, and the core argument of the case that we settled on after months of discussion was called the "takings argument," where the primary claim was that the TLC and the technology vendors, in bringing in the GPS system, were attempting to fundamentally alter the drivers' business practices and workday. As Bhairavi talked to drivers, part of what she was doing was trying to identify the best affidavits that could be filed for the case. At its heart it was an argument about autonomy. Rizwan, the Alliance's legal representative in the TLC court system, began keeping track of how the TLC was trying to use the GPS system in its enforcement procedures through a detailed study of the new cases coming to him that involved the automated trip datasheet. All OC members kept close tabs on what part of their daily business was paid by credit cards. The Alliance started a 1-800 hotline service that drivers could call and leave detailed messages on how their day had been affected by the new GPS system. I turned to reading and writing on the labor process as part of the effort to analyze the special significance of this technology in its relation to drivers' perceived sense of autonomy.

The most significant efforts at theorizing the labor process—from Marx and Engels (1994) to Braverman (1974)—grapple with the question of worker autonomy as capital organizes and reorganizes the labor process to facilitate the greater extraction of surplus value. The historic archetype of worker autonomy in Marx is the independent artisan. Marx describes in detail the processes by which the "independent" artisan loses autonomy—as capital takes control over the credit and commodity markets. The loss of autonomy is specific—the artisan is constrained inasmuch as he loses control over the cost of input and the price of outputs, but the core labor process—the process of production—where the artisan crafts and fashions his product, still remains fairly within the control of the artisan. This degree of loss of autonomy is what Marx refers to as the formal subsumption of labor to capital and the conditions under which absolute surplus value is extracted from the worker. It is not difficult to see the similarity between what Marx theorized as the condition of the nineteenth-century artisan/worker with the taxi industry, where a fixed lease is extracted from the driver and the ownership makes seemingly no attempt to interfere with the driver's workday.

For the nineteenth- and early twentieth-century capitalist, the extraction of absolute surplus value through formal subsumption was never the end of the road. Leaving the entire labor process still within the control of the worker/artisan meant that the capitalist could never maximize the extraction of surplus. It is precisely this constraint that leads the capitalist toward a different idea of subsuming labor—what Marx (1994) calls the real subsumption of labor. Here, we see a marked change—what several theorists note as the paradigmatic change in the history of capital-labor relations. With the emergence of the factory and the Fordist system of production, labor is brought into a relation of real subsumption viz capital. Here, the worker is fixed in time and space, production is transformed into a revolutionary new process of socialized production, and the extraction of surplus value is no longer absolute but relative. In other words, the capitalist seeks to control not just the external conditions of production, but also the labor process itself.

Surely, as I went back to read these theories of labor subsumption, for a moment it seemed that the taxi industry was a perfect example of formal subsumption, where capital stayed at the edges of the production process and attempted to extract the maximum. However, the more I spoke with other members of the OC, the further I got into the legal process on the "takings" case; the more I spoke with the mass base on their concerns and heard the messages on the 1–800 hotline, the more I was convinced that the sudden spate of anxiety around autonomy was emerging because something more subtle was happening.

As the October 1, 2007 deadline that marked the formal beginning of the implementation phase for GPS technology came closer, the level of activity in the Alliance office intensified again. At the injunction hearing scheduled for the last week of September, we were hoping that a large numbers of drivers would be in court. Once again, we were back on the M60. Once again, a new round of conversations, this time revolving around the specificity of the "takings" argument, were unfolding in the holding lots at LGA and JFK. And once again, as Bhairavi, Javaid, and I stepped onto the M60 two days before the injunction hearing, I reported yet another conversation.

"Shikdar was there again," I said, "and he is nonstop. Today he wanted to know if we could go back to some way of messing with the meter that could screw up the GPS. He kept saying, 'See, if the meter was not electronic, no

GPS could work.'" We laughed, and Bhairavi said that Shikdar is right, because one of the GPS vendors, Digital Dispatch, was forcing drivers who had bought the DDS system to change their meters. "Their system works only with one specific kind of meter," she said, "and so now the drivers are being forced to pay for new meters also."

As the ride drew to a close, I knew I was onto something, and I began to talk to Bhairavi and Javaid about formal and real subsumption. The more I talked, the clearer it became. The GPS system was so important for drivers because it was a technology that did not increase the level of formal subsumption, but instead actually engineered, in the subtlest of ways, an increasingly real subsumption-like character under the guise of formal subsumption. The introduction of the electronic meters in the mid-1980s had, I realized, given the ownership some degree of control, both in terms of information on driver earnings (so that the lease could be increased) and in terms of their capacity to program the meters to turn off at a specific hour on a specific day. Between the late 1980s until the years before the GPS, the TLC's enforcement rules expanded dramatically, forcing the drivers, for instance, to record more and more details about every fare they picked up, greater details about every break they took while on shift, and subjected them to an inspection of their cars every four months—a process that only got more and more technologized and detailed as the year went by.

All in all, the process in formation in the taxi industry since the 1980s has, with the GPS system, accelerated and extended beyond anybody's wildest imagination. It is not a simple case of either formal subsumption or real subsumption, but a complex hybrid of the two, tailored and crafted to the specifics of the taxi industry, that flexibly plays both ends of the margin between formal and real subsumption. This was what Surinder had been struggling with. At first glance nothing seems to have changed. Nobody was telling Surinder when to drive. Nobody was asking how much he was earning. Nobody was holding Surinder to a certain target figure of metered bookings. None of this was happening, and so it does seem like a case of formal subsumption. Until we begin to understand how the credit card system was placing all of Surinder's money in the hands of the broker, and that Surinder loses autonomy with every additional dollar charged on the credit card instead of being paid in cash. The appearance is of formal subsumption, but the structures of real subsumption are truly in place. It is a form of flexible subsumption that has been crafted for the taxi industry.

I call this flexible subsumption because its articulation not only uses, as in this taxi industry example, elements of both real and formal subsumption, but more important, because its uniqueness lies in the fact that similarly specific recombinations can be created for every industry. We have only to look around us to see the processes of combination at work. Whether it is security services, janitorial services, fast food franchises in the first world, call centers, or cotton farming using marginal farmers in India, we see the specific articulation of flexible subsumption crafted with the specificities of that industry in mind.

Collective Ethnographies and Material Theory

In the previous two sections, I attempted to render in as much detail as possible various aspects of the knowledge production process within an organizing project. My effort was to ensure some visibility to the nitty-gritty of knowledge production. Drivers as a body built an imaginary of the nature and scope of the technology that was soon to invade their taxis. The NYTWA, or the "organizers," did not lead this process of knowledge production, but were participants in its production. Later, smaller groups of drivers or even individuals like Surinder were able to distill some part of the scope of the system into sharper and more abstract forms of understanding. All such knowledge would recirculate back and forth between multiple spaces, including the Alliance office, and each round would be repackaged and would aid in the production of the next round of knowledge. Different actors from within the Alliance also played the role of reaching out, interfacing and bringing in certain forms of external knowledge and integrating it into the felt concerns and understandings being built on the ground. These include such aspects as the six-month process that helped build the lawsuit centered around the idea of "takings," and the ideas of flexible subsumption developed from Marx and Braverman aimed at not just understanding the consequences of the GPS system for the driver community, but also the broader changes within neoliberal economic formations.

The group outreach and mobilization process, conversations on the M60, the dynamic all-night discussions within the NYTWA office, the entry and exit of those external to the organization—lawyers, the Central Labor Council (CLC), technology experts from both academy and industry—the successive rounds of discussion within spaces such as airport lots and restaurants, at homes and in membership meetings, point to a complicated collective form of knowledge production, what I call "collective ethnography." The organizers traverse all of these different spaces and engage all of these different groups of people.

Identifying the work of the organizers as a collective ethnography allows for an interesting analysis of the ethnographic process. My point is not to place value immediately on the organizers' collective ethnography as better or worse than other ethnographies. Rather, conceptualizing the work of organizers as collective ethnography allows us to reflect on the material conditions of its production, and thus argue that all ethnographies are constituted and conditioned by their material spaces of production. Organizers formalize the knowledge that is emergent through these multiple levels, repackage and force each short cycle of knowledge production back into circulation, and facilitate the evaluation of the knowledge produced through external agents/allies. Thus, organizers facilitate the expansion of knowledge, and each round of knowledge is quite immediately returned to other levels for engagement. This knowledge is automatically implicated in the material reality of the worker community. The collective process ensures that all abstraction is constantly engaged with the materiality of the domain. It forces a short cycle of theorizations—and ensures that each round of theorization is immediately engaged with the materiality of the domain of organizing. In other words, even as my conversation with Surinder and Bhairavi in the Delta lot take me to Marx and flexible subsumption, it is immediately engaged back with the NYTWA's work—into conversations in the airport lots and restaurants, with lawyers, and, of course, the rest of the organizers. My own capacity as a theorist to hold onto the category of subsumption as relevant is determined in large part by this continuous cycle of engagement. The limits and parameters of my/our ethnography are set by the material conditions of the "work domain" in which I am engaged.

Thus far, I have tried to make intelligible the process of knowledge production within political work, and traced a line that allows us to see the same as a collective ethnography, thereby opening up the possibility of locating it as a different and valuable mode of knowledge production. To conclude this chapter, I wish to ask one final question: Beyond the fact that collective ethnography is the mode of knowledge production in political spaces by political actors, does it have any other value? It could be asked, for instance, how does it engage the external world outside of the immediate space of the political project itself (and the few specific external actors the project already engages)? Is it, we could ask, capable of engaging a broader community or society?

Political Work, Knowledge Production, and the Public Intellectual

The collective ethnography, I would argue, is a powerful point of departure for the work of a public intellectual. The primary reason for this is the very epistemic nature of a collective ethnography—a form of knowledge production that engages multiple groups of actors—often of significant diversity, both to generate the knowledge and then further to reengage the actors with other accumulations of collective knowledge. In other words, the collective ethnography starts out with the advantage of being "multilingual" inasmuch as it uses and translates the ideas and concerns of one group for another. This "multilinguality" of collective ethnographies already positions the knowledge thus produced as amenable for wider circulation through the aegis of a public intellectual. In sharp contrast to the popular multilinguality of knowledge produced in political spaces, ethnographic knowledge production out of academic spaces is marked by a monolingual and "parochial" knowledge form. Like all ethnographies, collective or academic, these also start out in community spaces and therefore inherit some of the multilingual quality. However, the academy, especially in the United States, has over the past three decades become an increasingly closed space with an internal market for symbolic material that has little or no value outside its borders. This dramatic transformation, so well captured in Russell Jacoby's *The Last Intellectuals* (2000) has effectively meant that the old tradition of academics who also served as public intellectuals is now merely a memory. Thus, while it is possible to explain the missing public intellectual from the academic space, how do we explain the almost complete absence of public intellectuals from within political spaces? Unlike the academic ethnography, the collective-organizer ethnography is, as we acknowledged, multilingual and ready for redistribution. While this question needs to be addressed systematically so that we can both put into place a cogent explanation and thereby begin the process of correcting the problem, I offer a hypothesis in closing. Political spaces do not produce public intellectuals because political actors do not believe that they produce any "valuable knowledge." I started this chapter with an acknowledgment that one of the most serious problems that besets the political world is an unfortunate and deep anti-intellectualism. In significant degree, owing to this strand of anti-intellectualism, most political actors do not believe that political spaces are just as valid and crucial spaces for knowledge production as is the academy. The identification of the idea of a collective ethnography, I hope, serves as a starting point for us to think through the urgent need to return to the project of active public knowledge production from deep within political spaces.

Notes

1. Garages control close to 25 percent of the medallions and represent the traditional ownership of the taxi industry. Brokerages are newer institutions that emerged in the 1980s from trading houses that managed medallions owned by third parties. Brokerage houses currently control over 50 percent of the medallions in New York City. While garages rent out a full taxi—car and medallion—together,

brokers lease out medallions to drivers, who then also enter a distinct car-buying contract facilitated by the broker.
2. RFI is now an integral part of government contracting procedures in New York city. When the city issues an RFI, it is essentially putting out a so-called open call for new ideas in a domain and/or information about firms that claim to have expertise in an area.
3. The OC is the New York Taxi Workers Alliance's main decision-making body, comprised of between ten and twenty drivers and one or two nondriver organizers. See http://www.nytwa.org for more on NYTWA.

References

Anderson, P. (2007). "A short history of the idea of the intellectual." Talk at Columbia University, New York, March 28.

Braverman, H. (1974). *Labor and monopoly capital: The degradation of work in the twentieth century.* New York: Monthly Review Press.

Chari, S. & H. Donner. (2008). "Ethnographies of activism." Conference at London School of Economics, February 29.

Jacoby, R. (2000). *The last intellectuals.* New York: Basic Books.

Marx, K. (1994). "The economic manuscripts of 1861–63." In *Collected Works*, ed. K. Marx & F. Engels, vol. 37. London: International.

Mathew, B. (2005/2008). *Taxi! Cabs and capitalism in New York City.* Ithaca, NY: Cornell University Press.

CHAPTER 11

Vanguards and Masses: Global Lessons from the Grenada Revolution

David Austin

Look is twenty years and the nation still hurting
People playing a waiting game, they just not talking
Is hard if men suffering on the hill for things they didn't do
People not relenting because they have their memories too
Dust don't disappear when you sweep it behind bed
People stay quiet but all the questions in their head
Is true time could heal and bad times could change people mind
But we have to figure how to talk, leave the hurt behind
…
—*Merle Collins, "Shame Bush"* (2003)

Revolutions do not happen outside of you, they happen in the vein, they change you and you change yourself, you wake up in the morning changing. You say this is the human being I want to be. You are making yourself for the future, and you do not even know the extent of it when you begin but you have a hint, a taste in your throat of the warm elixir of the possible.
—*Dionne Brand* (1994)

Culture and politics are one. Before the revolution, the dictatorship used to project that culture must stay out of politics, and yet they had used politics to try to stifle culture! But that's impossible! If you say culture should stay out of politics, you are saying that culture should stay out of the people, and then you don't have culture.

—*Flying Turkey* (1984)

Introduction

On March 13, 1979, the New Jewel Movement (NJM) seized power in Grenada in a virtually bloodless takeover by the National Liberation Army, a clandestine military force within NJM (Meeks, 1993). These events were met with widespread support and enthusiasm in Grenada. Eric Gairy, the overthrown prime minister, was a former labor leader who played the lead role in Grenada's independence movement. But as A.W. Singham (1968) eerily illustrated in his often neglected study of nationalism, *The Hero and the Crowd in a Colonial Polity*, Gairy's penchant for power and control was apparent quite early in his political career, and in retrospect, his repressive, authoritarian rule was perhaps predictable. Gairy used political intimidation, corruption, electoral fraud, and his much-feared paramilitary force, the Mongoose Gang, to strike terror in, suppress, imprison, and murder his opponents, while internationally aligning himself with political pariahs such as Chile's Augusto Pinochet.

The Grenada Revolution ended Gairy's reign, and between March 1979 and October 1983, this tiny Caribbean island engaged in a revolutionary process that inspired individuals and movements across the globe. It was a short-lived process that traumatically imploded in a manner even more dramatic than its ascendance on the world stage, due in part to inexperience and the pressures inherent in attempting to carry out a revolution in the United States' "backyard." But brief as it was, the Grenada Revolution was no fleeting moment. In this chapter I will demonstrate that this revolution embodied in microcosm many of the challenges that confronted twentieth-century socialism and social movements, and that, as such, Grenada's experience has profound and universal implications for our understanding of social transformation and the dynamics of liberation.

I will suggest that without ignoring the role played by the pressures of the Cold War and the threat of U.S. intervention in Grenada, the revolutionary leadership's messianic faith in Marxist-Leninist theory and its blind dependence on vanguard politics distanced it from the majority of Grenadians, profoundly shaping the direction that the revolution would ultimately take.

Lastly, I will argue that if existing and future social movements and socialist governments can learn from this revolution's inspiring but tragic experience, then perhaps they will avoid many of the pitfalls that beset Grenada. Lessons from the Grenada Revolution are not only important to the revival of the Caribbean left. Grenada was a microcosm of the world in which Cold War politics played themselves out with both inspiring and tragic consequences, and the world, particularly the generic left, has much to learn from Grenada's experience.

The Grenada Revolution

1979 was an important year for the global left. In Iran, the unpopular monarch, Mohammad Reza Shah Pahlavi, was overthrown by a popular movement. The Sandinistas overthrew the corrupt Anastasio Somoza regime, promising a new day in Nicaragua. The Cuban Revolution remained an important symbol of resistance to U.S. imperialism. In the Anglophone Caribbean, popular movements and left political parties—the Working People's Alliance in Guyana, Youlou United Liberation Movement in St. Vincent and the Grenadines, the Antiguan Caribbean Liberation Movement, and the Workers Party of Jamaica, to name a few—offered hope of a genuine break with colonialism in the region.[1] And while support for prime minister Michael Manley's democratic socialism began to wane in Jamaica under the pressure of constant social and economic destabilization from the United States, Manley himself was seen as important international symbol of resistance to neocolonialism and underdevelopment.[2]

But more than any other political development of that period, Grenada represented a genuine David and Goliath story, in which the tiny island confronted imperialism, refusing to be a playground in the United States' backyard. Under the People's Revolutionary Government (PRG) of Grenada, institutions such as the National Women's Organization, the National Youth Organization, and the Parish and Zonal Councils quickly assumed prominent roles in the country. Education, health care, and housing, all woefully neglected under Gairy's regime, became top priorities for the PRG (Pasley, 1999). Literature and the arts in general also flourished during the revolution as poets and calypsonians exercised their newfound ability to freely express themselves, and eliminating illiteracy on the island became a top priority of the new government (Pasley, 1999; Searle, 1984). But despite the tremendous social, economic, and cultural gains made during the Grenada Revolution's brief existence, Brian Meeks (1993), one of the Caribbean left's foremost theorists, has described it as "a revolution from above, marginally distinguishable from a coup d'état by its execution by armed irregulars and by the willingness of the leadership to mobilise popular support, though firmly under its command" (p. 156).

Within four years of its inception, the revolution imploded. A dispute over the Central Committee's proposal for joint leadership involving the country's principal leaders, prime minister Maurice Bishop and deputy prime minister Bernard Coard, led to the executions of Bishop and several other prominent government figures.[3] The executions, the subsequent U.S. invasion of the country, and the imprisonment of Bernard Coard and several government and military figures[4] marked the end of Grenada's short-lived revolutionary process.

The collapse of the Grenada Revolution also signaled the demise of the organized Caribbean left (Kamugisha & Trotz, 2007; Levitt, cited in Austin, 2007) and resulted in a double silencing of the revolution itself. The implosion, subsequent U.S. invasion, and the resulting trauma effectively silenced the revolution. Open discussion about the revolution was also silenced, both in Grenada and among West Indians who were closely associated or who worked in solidarity with it. According to Grenadian poet and novelist Merle Collins, Grenadians have yet "to figure how to talk, leave the hurt behind," resulting in what she describes as the "silence people keeping" (Scott, 2007, p. vi).

But there is evidence that the blanket of silence is lifting and that the revolution's open wounds are beginning to heal. The February 2007 issue of *Small Axe: A Caribbean Journal of Criticism*, a leading journal on the Caribbean, was dedicated to Grenada and contained some intriguing articles, including one by Nicole Phillip, which I discuss below. A series of events in Toronto in autumn 2008, "Memory and Renewal: 25 Years After the Invasion: The Grenada Story," largely sponsored by the Caribbean Studies Department at the University of Toronto, provided a forum for younger and older Grenadians and scholars to discuss its significance, in many cases for the first time publicly. This event was followed in spring 2009 by a conference, "Remembering the Future: The Legacies of Radical Politics in the Caribbean," organized by the Center for Latin American and Caribbean Studies, Pittsburgh University, and two months later, two panels at the Caribbean Studies Association's annual conference in Jamaica focused on the legacy of the revolution. With the recent release of Bernard Coard, among others involved in the revolution, from prison in Grenada, discussion and debate about the revolution will no doubt intensify in the Caribbean.

Rise and Fall

The Grenada Revolution was the culmination of the wave of left and Black Power movements that emerged in the aftermath of the expulsion of Walter

Rodney[5] from Jamaica in October 1968 (Meeks, 1996). The Anglophone Caribbean has a strong tradition of militant struggle, including Indigenous and black slave rebellions, the labor rebellions that gripped the region in the 1930s, and a tradition of socialist and Marxist-Leninist politics (Mars, 1998). But according to historian Robert Hill (Scott, 1999), the Caribbean New Left "never offered a cogent programme of change," the result being that "when the popular enthusiasm of protest died down, it [the New Left] quickly wilted" (p. 105).

But there were exceptions to Hill's rule. Grenadian Franklyn Harvey was one of the driving forces behind two New Left groups—the Trinidad and Canada-based New Beginning Movement (NBM), and Movement for the Assemblies of the People (MAP) in Grenada—both of which theorized and attempted to lay roots for social transformation in the Caribbean (Henry, 1992; Mars, 1998). Little has been written about NBM and MAP, whose work was ineluctably tied. Harvey was a student of C.L.R. James and a member of the Caribbean Conference Committee and the C.L.R. James Study Circle in Canada along with Hill. Both groups engaged in deep reflection on social change in the Caribbean (Austin, 2009). His study and James's influence is evident in Harvey's NBM booklet, *Rise and Fall of Party Politics in Trinidad and Tobago* (1974), which can be read as part of MAP's theoretical lineage.

Harvey betrays his ties to C.L.R. James from the outset in *Rise and Fall*, arguing that "[t]he objective tendency in Trinidad and Tobago today, and in the Caribbean as a whole, is towards the self-organisation of workers, farmers, unemployed, students, etc. to fight their own struggles" (p. ii). His statement echoes his reflections on the 1968 Paris uprising in *Caribbean International Opinion* (1968). Drawing on James's *Notes on Dialectics* and *Facing Reality*, Harvey identified the action committees that emerged in France as a new form of social organization that would permit workers to independently manage production and govern society (Austin, 2009). In *Rise and Fall*, Harvey extends his notion of self-organization to the Caribbean, arguing that both political parties and trade unions are outmoded and that "mass organizations of workers and farmers at places of work and in the villages and communities must be the basis for the control of state power" (p. ii) in what he describes as "Revolutionary Democracy" (p. 6). The entire booklet is an elaboration of this idea. Harvey eschews preordained models, suggesting that the "working-masses" will develop the precise form of social organization in the process of struggle. This will emerge "in the factories; on the estates; in the industries; in the department, on the bloc; in the Village" (p. 70), coming together to form a national mass organization. Interestingly, at this stage Harvey does not dismiss the significance of what he describes as "a revolutionary vanguard" in social movements, although he cautions—and this is of significance for the Grenada Revolution—that the revolutionary vanguard today *cannot* be a vanguard whose perspective is the seizure of state power *with the support of* the working class to wield it *in the interest of* the working class. Yet he argues that a revolutionary vanguard's priority must be seizure of power by "*the Mass organization of the working-class as a whole* and the establishment of *working-class social power* in all spheres of activity," while calling for the "abolition, the destruction of all vanguard organizations; *not after* the revolution but *in the very process of making the revolution* (pp. 68–69, emphasis in original).

Although we are still hovering in the realm of theory here, Harvey offers no sense of how the vanguard will self-destruct in the process of revolutionary struggle. In another apparent contradiction in his notion of self-organization, he argues that it is the vanguard that gives direction to the self-organizing mass: "The concrete task of the revolutionary vanguard in Trinidad and Tobago today is to tell the working-class by agitation and propaganda that it must seize state power with its mass organizations of the whole class *to the last individual*" (p. 69, emphasis in original). Here we confront a dilemma: If the revolutionary vanguard is to lead and instruct the working class, what does self-organization mean? If the vanguard is supposed to abolish itself in the process of seizing power, how and when does this occur? In other words, despite the language of self-organization, mass organizations, and so on, leadership and vanguardism were crucial to Harvey's notion of social transformation, and he offers no theoretical or practical method or approach to checking the kind of elitism and lack of accountability that characteristically take hold once leaders become entrenched. This is a phantom that haunted the global left with grave consequences for the Grenada Revolution once it made the transition from a revolutionary vanguard party to government.

New Jewel and Leninism

In the early 1970s, Harvey attempted to apply his Jamesian approach to Grenada. He cofounded and was one of the intellectual forces behind MAP (Marable, 1987), which promoted rule through village and popular assemblies (Meeks, 1993) and bore the marks of James's decentralized, bottom-up approach to politics (Henry, 1992; Meeks, 1993). He began meeting with people throughout the country, discovering that Grenadians were ready for alternatives to Gairy's rule. MAP eventually merged with the Joint Endeavor for Welfare, Education and Liberation (JEWEL) and other groups to form the NJM. Village assemblies eventually became a component part of the

NJM's platform and program,[6] although, as Bernard Coard (1989) has acknowledged, assemblies emerged "accidentally" (p. 2) during the revolution when large numbers of workers showed up uninvited to NJM Party delegates' council meetings, transforming them into the revolution's first assemblies. In other words, workers inaugurated the assemblies and dragged the PRG into the process. As Coard laments, had Grenada's leadership taken the dispute over joint leadership to the various assemblies—women, workers, youth, etc.—in 1983, the outcome of the joint leadership discussion might have been different.

Largely inspired by Harvey's perspective, the NJM's early platform shunned the idea of a single leader, often referred to as one-man-ism in the Caribbean, and was hostile to elections, but embraced notions of African socialism and "the Marxian/Rousseauist view that deeper, participatory forms of democracy were superior" to parliamentary politics (Meeks, 1993, p. 146) Drawing on C.L.R. James, it also emphasized that the state should be built from the bottom upward and expressed hostility to official communism, instead stressing spontaneous uprising.

Although James's direct influence on the NJM is evident from its inception, Marable (1987) suggests that at this point "Bishop himself had a vague understanding of James's Marxism, but probably had done little serious study on his own" (p. 208). On the other hand, Coard, of the Organization for Research, Education, and Liberation (OREL), was closely associated with the Worker's Liberation League, eventually called Worker's Party Jamaica, which was closely allied with the Soviet Union, where vanguard politics was almost sacred. Harvey, who opposed the transformation of the NJM into a vanguard political party, relocated to Canada shortly after its formation, periodically returning to confer with Maurice Bishop and other leaders of the revolution.

The early influence of James and the emphasis on a decentralized notion of power begs the question: Why did vanguardism become entrenched in the NJM? The NJM turned to vanguardism after several unsuccessful attempts at unseating Gairy by electoral means. Meeks (1993) cites a 1975 essay by Coard: "This failure to organise the working class as an independent force with its own tactics and acting on its own behalf, was to lead at a most critical period in the revolutionary struggle, to a NJM finding itself *objectively tailing* the bourgeoisie" (p. 149, emphasis in original). The Leninist party structure of tightly organized cadres dedicated to removing Gairy from power was seen as the antidote to the NJM's inability to win real political concessions from the Gairy regime. By 1974 the NJM began to reorganize itself along Marxist-Leninist lines in order to face the challenges of an increasingly desperate and repressive Gairy government (Meeks, 1993). The vanguard party's tight organizational structure, secrecy, and emphasis on loyalty and strict adherence to party dictates helped shield NJM members from Gairy's repressive tactics and the brutality of his paramilitary force. Moreover, this shift in the direction of Marxist-Leninism was consistent with a global political trend, and with a few notable exceptions, including Trinidad's labor movement and the Antiguan Caribbean Liberation Movement (ACLM) under Tim Hector's leadership,[7] most left movements and parties in the Caribbean and Latin America functioned along Marxist-Leninist lines (Meeks, 1993).

But given the harsh and repressive conditions and the political stakes in Grenada, a strong argument could be made in favor of a tightly knit, centralized political organization in which discipline and loyalty were paramount. Yet this approach was fraught with problems once the NJM came to power, and, as the collapse of the Grenada Revolution itself demonstrated, is antithetical to the kind of popular participation in all spheres of social life that is ultimately the basis for meaningful social transformation.

Gendering Revolution

The role played by women and women's groups during the Grenada Revolution is instructive, not only in terms of gender dynamics during the revolution,[8] but also because it underscores fundamental issues related to leadership and popular participation. In other words, in addition to the significance of gender on its own terms, examining gender dynamics in Grenada provides us with insight into the ways in which power dynamics in general played out during the revolution. Nicole Phillip (2007) describes how women's groups, including those opposed to Gairy's regime, were disbanded or absorbed into the PRG's National Women's Organization (NWO). For the PRG, autonomous women's groups were incompatible with the aims of a socialist state, which meant that the Progressive Women's Association (PWA), one of the groups that had opposed Gairy's government, was not permitted to establish a women's reading center for fear of promoting "opportunists and CIA elements bringing revisionist Maoist and Trotskyist literature" (Coard, cited in Phillip, 2007, pp. 41, 42), a remark that highlights the extent to which intrasocialist invective had penetrated the revolution.

Led by Phyllis Coard, the NWO played an important role in the revolution. It distributed cooking oil and dried milk to those in need, established cooperatives and created employment, and through the NWO, women helped to build roads, community centers, schools, and preprimary nurseries (Phillip, 2007). The PRG took steps to attack machismo and patriarchal practices, despite considerable resistance, including from senior women who did not believe that younger women should be entitled to maternity benefits that they themselves had not been privy to. Yet sexism remained rampant and women were expected to carry out party work while maintaining their traditional family roles. Phillip writes:

> [I]t can be argued that in spite of its socialist revolution, the positions of power held by women in Grenada were of the "kitchen cabinet" type [insofar as] Grenadian women were placed at the head of such ministries as education and women's affairs. Thus while the revolution may have increased the number of women in positions of power, it did not revolutionize the *type* of positions they held. The woman as minister was limited to her role as social worker and teacher. (p. 65, emphasis in original)

Moreover, "the concern of the revolutionary leadership to end women's confinement to traditional roles too often seemed limited to making their labour available to the regime. Women became as free as men to work outside the home while men remained free from work *within* it" (p. 66, emphasis in original). In other words, despite its laudable achievements, the NWO and PRG were unable to fundamentally change Grenada's gender dynamics, although it must be emphasized that the revolution lasted only four years, a short period of time in political terms.

At the root of these problems was the fact that the NWO was not an autonomous body, but what Phillip describes as a "tool of the PRG" (p. 45). This perhaps sounds self-evident given that the NWO was established by the PRG, but only if we assume that the state, socialist or otherwise, is the sole and legitimate representative of the people. Like the revolution's other institutions, the NWO was part of and accountable to the state apparatus. In theory, this might have been justifiable on the grounds that the PRG led the revolution, but in practice this "winner gets the spoils" mentality proved disastrous. This is precisely the point that Meeks (1993) makes about the PRG's overall approach: Unless this rule from above "was converted into a more organic and democratic movement of the people, [it] possessed the same inherent fragility as the Gairyite state, derived from a shattered alliance at the top, with similar potential for crisis when placed under pressure" (p. 157).

The revolution reached a breaking point during a discussion within the PRG's Central Committee of a proposal for the joint leadership of Maurice Bishop and Bernard Coard.[9] At this stage, a perilous form of ideological rigidity was rampant within the Central Committee, one that suggested that if a political idea or concept was not culled from Soviet Marxist-Leninist textbooks, it was of little value. Yet, in the eyes of some Grenadians, Bishop's refusal to accept joint leadership was unforgivable.

Nicole Phillip's article represents one of the rare occasions in which Grenadians involved in the revolution are cited criticizing Bishop on this score, while at the same time severely reproaching the PRG for its doctrinaire stance, which laid the groundwork for the events that culminated in his execution (Phillip, 2007).

Vanguards and Masses

What can we conclude and learn from the rise and demise of the Grenada Revolution? How does it inform our understanding of the prospects and challenges of social transformation in the Caribbean today, and global dynamics of liberation in general? The question of vanguard parties and leadership is crucial here. Given the gargantuan challenges that confronted the PRG, including U.S. efforts to destroy the revolution, it is not difficult to understand how Grenada's leadership could have become so detached from the pulse of ordinary Grenadians and why it buckled under the weight of internal and external pressures. But the way in which the leadership of this small and close-knit society of 100,000 people callously turned against itself seems hard to fathom. It is impossible to divorce the gulf that developed between the PRG, and particularly its Central Committee, and the population from the "science" of Marxist-Leninism that was grafted from the USSR and practiced in Grenada. It was a practice that, less than ten years after the Grenada Revolution's collapse, Russians would themselves reject. This brand of Marxism was antithetical to popular participation and genuine democracy, a point amply demonstrated by the way in which the majority of Grenadians increasingly became spectators in their own revolution.

For Richard Hart, an elder statesman of the Caribbean left and a Jamaican who served as Grenada's attorney general during the revolution, inflexible leadership and not Soviet-style vanguard politics were at the root of the Grenada Revolution's implosion. In an introduction to a collection of Bishop's speeches, Hart argues:

> The existence of a Leninist vanguard party to provide the necessary orientation at such historic moments is vital for the achievement of popular power. Without such leadership... [and] in times when unrest and dissatisfaction have increased beyond the level of popular tolerance, blind uncoordinated outbreaks of disorder can easily occur which are then suppressed by the armed forces with purposeless loss of life and destruction of property. Such directionless upheavals have indeed occurred over the past half century in other parts of the English-speaking Caribbean area. (Hart, 1984, pp. xiii–xiv)

Hart argues that the NJM demonstrated exemplary leadership prior to the revolution, but is critical of Bishop's attempt to turn back a decision that the Central Committee had already agreed upon. He also suggests that, having concluded that the Central Committee was not willing to reopen the discussion on joint leadership, Bishop decided to go to the masses with the issue, over the heads of the Central Committee members and the party, a strategy that Mao Tse-Tung used when he lost control of the Central Committee of the Chinese Communist Party (Hart, 1984). Hart also mentions a rumor that Bishop is said to have spread, accusing Coard and his wife of plotting to kill him. Ultimately, Hart places blame squarely on the shoulders of the revolution's entire leadership. He suggests that Bishop's volte-face on joint leadership violated the principles of democratic centralism—a commitment to accept party decisions—to which the NJM adhered, and that the resulting disappointment of his fellow party members was understandable. Having said that, Hart (1984) argues that "the Marxist approach requires that all the circumstances of a given situation be taken into account" (p. xxxvi).

But what exactly is the Marxist approach that Hart refers to? Or, posed another way, whose Marxism? Clearly, competing Marxisms were at play in the Caribbean. The version of Marxism that dominated Grenada appeared to leave little room for the kind of critical reflection—dialectical reasoning—that Hart suggests was necessary. "Given Bishop's determination not to implement the joint leadership decision" (Hart, 1984, p. xxxvii), to which he was bound by the very practice of democratic centralism, to which he adhered as leader of the revolution, and given the immense popularity of Bishop, Hart adds:

> [T]he prudent course for the Central Committee to adopt...was to explore other options for improving the structure and quality of the leadership of the party until a compromise formula had been found which both Bishop and other members of the Central Committee could accept.... Any other course was bound to bring the party into conflict with the masses. (p. xxxvii)

In hindsight, this is certainly true, but what is absent here is an analysis of how the inability to reach an alternative was embedded in the "brand" of Marxism that came to dominate the PRG, culled from the Soviet Union, where popular dialogue and accountability to the population were largely absent.

Although the PRG encouraged a degree of popular participation, the notion of self-organization appears to have disappeared entirely from its program and practice. The idea of popular assemblies and government *led from below* had become a distant memory. The image of Grenadians parodying passages of the "science" of Marxism-Leninism is one of the sad, unfortunate, and telling spectacles of Grenada's revolutionary process. Those who possessed the "science" were set apart, or set themselves apart, from those who did not, which ultimately put those who did in a position of privilege through which they could "manners," that is to say discipline, those who transgressed the party line.

Conclusion

Because of the revolution's experience, both with communism and the success of the revolution's social programs, Grenada is ideally situated to teach us about the challenges of revolutionary politics. Given its small size, it serves as a kind of political laboratory, though in saying this, I in no way mean to diminish the experiences of Grenadians between 1979 and 1983, or to reduce their experiences to a kind of social "experiment" divorced from their lived triumphs and trauma. It is precisely the "scientific" approach to politics that led to the revolution's callous destruction. What I am suggesting is that far too often, the Caribbean is seen as a playground for tourists and foreign interests, and not a place where people live, breathe, create, and struggle, and it is a place whose struggles inform our understanding of the dynamics and challenges inherent in social transformation.

Grenada warns us that any process of social transformation must be true to the autochthonous culture, spirit, and social mores of the society involved, otherwise it is destined to run into conflict with itself. In 2002, writer George Lamming, who for almost fifty years has been one of the Caribbean's preeminent voices of reason, described what he calls the "colonized left," middle-class leaders who had the benefit of colonial education, but who failed to expunge its negative effects; leaders who uncritically imposed foreign models and texts on Caribbean people while failing to draw on the contemporary and historical realities and the culture of the people (Scott, 2002). This left read Marx's writing but failed to translate "the essence and spirit of meaning of that text into a wider collective experience" (p. 176). Tim Hector (1998) has similarly argued that "the Grenada Revolutionary collapse reminds [us] that we need to return to our own historic and historical experiences, or there will be the persistent inauthenticity in which we all live in this region" (p. 58). The idea of drawing political lessons from the Caribbean's historical experience as opposed to the wholesale adoption of foreign political models is also echoed by Rupert Lewis, a former member of the Workers Party of Jamaica (Scott, 2001). The relationship between ideas and action is undeniable, and knowledge, regardless of its origins, often finds roots among foreign hosts. But knowledge and doctrine are not the same species, and genuine dialectical materialism, presumably at the core of Marxist-Leninism, is antithetical to the doctrinal stances and inflexible categories of thought (James, 1980).

Clearly, both the hard-nosed, doctrinaire brand of Marxism adopted in Grenada and U.S. efforts to destabilize and ultimately snuff out the Grenada Revolution are part of a history in the Caribbean that warrants revisiting with fresh eyes. Such a revision would involve the question of leadership, and more specifically the relationship between those who lead and those who are led. Vanguards of various kinds characterized the Caribbean left in the 1970s and 1980s, and as we learned, even Harvey (1974) was not completely dismissive of vanguard parties. Here the issue is not so much whether leadership is necessary to the success of social movements, but the nature of that leadership as well as the way in which—or whether—a vanguard party can make the transition to civilian rule once in power. As Richard Hart indicates above, notions of self-organization and spontaneity, which suggest that an amorphous "mass" will simply rise up and spontaneously take power, have obvious weaknesses. What does organized spontaneity mean? The apparent contradiction inherent in the notion is obvious, but this is precisely what both Harvey and Hector, drawing on James, argue in an attempt to escape the pitfalls of centralized government and vanguard leadership.

In a recent reflection of the legacy of the Grenada Revolution and the Caribbean left, Rupert Roopnaraine (2010), once a prominent figure in Guyana's Working People's Alliance, called for a revision of the left's attitude toward parliamentary politics, organizing in order to "launch a people's challenge to the old politics and its entrenched interests, armed with a radical and humane social, economic and political reform agenda, with the dismantling of the ossified state system as the number one priority" (p. 34). This alone, he argues, can bring the Caribbean closer to the ideals of freedom that the best of the Grenada Revolution embodied.

Roopnaraine's point on electoral politics is instructive, although he appears to overlook the circumstances that made electoral politics impossible in Grenada under Gairy. Yet the wounds of colonialism and the Cold War still fester in the Caribbean, and for Roopnaraine (2010), the revolution's mortal sin was to leave unfulfilled "the promise of the ordinary people taking direct control of their destiny, devising the strategies of their own development, holding those who would be leaders and representatives to account" (p. 32). The Grenada Revolution's collapse

> delegitimized vanguardism and instructs that it is more imperative than ever to build alliances across differences. We must learn the strength

of otherness and difference, and, above all, we must guard against the fatal arrogance of righteousness. We must focus on what we agree on and work persistently and with integrity on differences that are primary contradictions. The community-based organizations outside the political parties are already networking for health, literacy, child protection, environmental sanity, and disaster preparedness, and against family violence and rape. We must deepen the work, and multiply and connect these networks within and across territories. (Roopnaraine, 2010, p. 33)

Here we would perhaps do well to consider what Frantz Fanon described in *The Wretched of the Earth* (1968) as "a mutual current of enlightenment and enrichment" (p. 143). In this process, political intellectuals recognize the acquired knowledge of "ordinary" citizens as the two foster a relationship of mutual recognition, respect, and collaboration, much of which can only be achieved once both parties have engaged in a joint process of learning and political engagement. One form of leadership, and certainly not that of the middle class—left, conservative, or in between—is not privileged over the other. Fanon seems to capture the essence and spirit of what leadership should entail, a symbiotic process in which the "mass" and the leadership feed and fuse into one another with the goal of making them indistinguishable, a notion that was sorely missing in Grenada when its need was most dire.

During the Grenada Revolution, literacy made tremendous gains, and the arts, and particularly literature, flourished. It is therefore fitting to end this chapter where it began, with a passage from a Grenadian and one of the Caribbean's most compelling novelists and poets, Merle Collins. In a 2005 interview, Collins discussed the significance and legacy of the Grenada Revolution:

To me when I do all of that recreating of Grenada and of the Grenada space and of the experiences and so on, I am understanding not only myself and other Grenadians, but humanity. Grenada is where a small portion of humanity that's responsible for my socialization, is centered. And that is of such importance for me personally, for me in the world, for all of those people there and for an understanding of self and society. (Bishop & McLean, 2005, p. 64)

"I think of that little country Grenada—all of those lessons ever learned there—I think those have been huge lessons for a lot of people worldwide" (p. 65). This, perhaps far more than the remarkable gains that were made in Grenada between 1979 and 1983, and despite its tragic end, is a gift from Grenada to the world. If the Caribbean and humanity as a whole can learn from Grenada's experience, the Grenada Revolution will not have been in vain.

Notes

1. For a detailed study of the Caribbean left during this period, see Perry Mars (1998).
2. For an example of Manley's role as advocate for the third world, see Manley (1983).
3. Those executed included Jacqueline Creft, Maurice Bishop, Unison Whiteman, Norris Bain, Fitzroy Bain, Keith Hayling, Evelyn Grant Bullen, and Evelyn Maitland.
4. Those imprisoned included Bernard Coard, Hudson Austin, Christopher Stroude, Liam James, Leon Cornwall, Selwyn Strachan, Phyllis Coard, John Ventour, Ewart Layne, Colville McBarnette, Dave Bartholemew, Lester Readhead, Cecile Prime, Cosmos Richardson, Andy Mitchell, Vincent Joseph, and Callistus Bernard.
5. Walter Rodney was a historian and Pan-Africanist from Guyana, best known for his book *How Europe Underdeveloped Africa* (1973). In October 1968 he was expelled from Jamaica, where he taught at the University of the West Indies, by Jamaica's government for his involvement in alleged subversive activities.
6. I am grateful to Franklyn Harvey for several conversations with him between August and October 2002 on the formation of the NJM.
7. Paget Henry (1992) asserts that had the ACLM's political approach been applied during the Grenada Revolution, it might have averted the sudden collapse in 1983.
8. For an overview of the role of women in Caribbean societies in the 1980s, see Ellis (1986).
9. The issue of joint leadership was raised in the Central Committee as popular support for the revolution and its economic fortunes began to decline. The Central Committee agreed that the combined leadership of Bishop and Coard would help bolster the revolution at a critical stage in its development. Bishop initially agreed with the decision, but later changed his mind.

References

Austin, D. (2007). Development, change and society: An interview with Kari Levitt. *Race and Class* 49(2), 1–19.

———. (2009). An embarrassment of omissions, or rewriting the sixties: The case of the Caribbean conference committee, Canada, and the global new left. In *New world coming: The sixties and the shaping of global consciousness*, ed. K. Dubinsky, C. Krull, S. Lord, S. Mills, & S. Rutherford, pp. 368–378. Toronto: Between the Lines.

Bishop, J. & D. N. McLean. (2005). Working out Grenada: An interview with Merle Collins. *Calabash: A Journal of Caribbean Arts and Criticism 1*(2), pp. 53–65.

Brand, D. (1994) *Bread out of stone: Recollections sex, recognitions race, dreaming politics.* Toronto: Coach House Press.

Coard, B. (1989). *Grenada: Village and workers, women, farmers and youth assemblies during the Grenada revolution: Their genesis, evolution and significance.* London: Caribbean Labour Solidarity and the New Jewel Movement in conjunction with Kari Press.

Collins, M. *Lady in a Boat* (London: Peepal Tree Press, 2003), p. 51.

Ellis, P. (ed.) (1986). *Women of the Caribbean.* London: Zed Books.

Flying Turkey (1984). Interview in Searle, C. *Words Unchained: Language & Revolution in Grenada*, p. 222.

Fanon, F. (1968) *The wretched of the earth.* New York: Grove Press.

Hart, R. (1984). Introduction. In *In nobody's backyard: Maurice Bishop speeches*, ed. C. Searle, *1979 to 1973*, pp. xi–xli. London: Zed Books.

Harvey, F. (1974). *Rise and fall of party politics in Trinidad and Tobago.* Toronto: New Beginning Movement.

Hector, T. (1998). Pan-Africanism, West Indies cricket, and Viv Richards. In *A spirit of dominance: Cricket and nationalism in the West Indies*, ed. H. M. Beckles, pp. 45–68. Barbados: Canoe Press University of the West Indies.

Henry, P. (1992). C.L.R. James and the Antiguan left. In *C.L.R. James's Caribbean*, ed. P. Henry & P. Buhle, pp. 225–262. Durham, NC: Duke University Press.

James, C. L. R. (1980). *Notes on dialectics: Hegel, Marx, Lenin.* London: Allison and Busby.

Kamugisha, A., & A. Trotz. (2007). Editorial. *Race and Class 49*(2), i–iv.

Manley, M. (1983, December). We are told…we must have more of the disease. *Harper's Magazine*, 24–28.

Marable, M. (1987). *African and Caribbean politics: From Kwame Nkrumah to Maurice Bishop.* London: Verso.

Mars, P. (1998). *Ideology and change: The transformation of the Caribbean left.* Detroit: Wayne State University Press.

Meeks, B. (1993). Meeks, *Caribbean revolutions and revolutionary theory: An assessment of Cuba, Nicaragua, and Grenada*. London: Macmillan Press.

———. (1996). *Radical Caribbean: From black power to Abu Bakr.* Kingston, Jamaica: Press of the University of the West Indies.

Pasley, V. (1999). The social and cultural record of the Grenada revolution, 1979–1983. Paper presented at 31st annual meeting of the association of Caribbean historians, Havana, Cuba, 1999, pp. 1–12.

Phillip, N. (2007). Women in the Grenada revolution, 1979–1983. *Small Axe 11*(1), 39–66.

Roopnaraine, R. (2010). Resonances of revolution: Grenada, Suriname, Guyana. Interventions: International Journal of Postcolonial Studies 12(1), 11–34.

Scott, D. (1999). The archaeology of black memory: An interview with Robert A. Hill. *Small Axe 3*(1), 80–150.

Scott, D. (2001). The dialectic of defeat: An interview with Rupert Lewis. *Small Axe* 5(2), 85–177.

———. (2002). The sovereignty of the imagination: An interview with George Lamming. *Small Axe* 12(2), 72–200.

———. (2007). Preface: the silence people keeping. *Small Axe*. 11(1), vi.

Searle, C. (1984). *Words unchained: Language and revolution in Grenada*. London: Zed Books.

Singham, A. W. (1968). *The hero and the crowd in a colonial polity*. New Haven, CT: Yale University Press.

PART III

Making Knowledge and Learning from Peasant and Indigenous Peoples' Struggles

CHAPTER 12

Learning and Knowledge Production in Dalit Social Movements in Rural India

Kumar Prasant and Dip Kapoor

Introduction

Even in these modern times, in South Orissa as in India, the Dalits [meaning "downtrodden"/outcastes] are still treated as untouchables by the elite and ordinary people. Having undergone three thousand years of slavery and discrimination, the Dalits find it nearly impossible to get out of this terrible trauma. Even if untouchability was abolished under India's Constitution in 1950, the practice remains very much a part of Indian life. In the upper caste dominated conservative villages and semi-urban centers untouchables are not supposed to move freely in the streets occupied by higher castes. They are not allowed to use the same wells, visit temples, drink from the same cups in tea stalls, or lay claim to land that is legally theirs. Dalit children are frequently made to sit in the back rows in classrooms or outside in the scorching sun in 40 degree centigrade temperatures, and as a whole Dalits are still made to perform degrading caste-customary practices and rituals in the name of caste.... Any attempt by Dalits to assert their rights is usually met with violent consequences with maximum brutality and degradation including being forced to drink cow urine, public floggings, decapitation, gang-rape, murder....

(Dalit leader, Adivasi-Dalit Ekta Abhijan movement, interview notes, January 2008)

The genesis of the Indian caste system can be traced back to 1500 BC and the ancient Hindu Vedic scriptures, wherein the sixth and tenth *Mandalas* of the *Purushasukta* hymn in the Rig Veda defines the four *varnas* as *Brahmanas* who came from his mouth (keepers of sacred knowledge and rituals), *Kshatriyas* from his arms (warriors and protectors), *Vaishyas* from his thighs (traders and farmers), and *Shudras* who came from his feet (menial labor). Dalits had no place in this religio-social scheme and are referred to as *Avarna* (outside/outcaste), *Dasa*, and *Dasyus* (servants), relegated to performing polluted and polluting tasks such as sewage disposal, tanning of hides, and the removal of carrion and refuse. The *Chandogya Upanishads* (800–600 BC) clarifies this scheme further and refers to the said castes, and also compares *Chandalas* (outcastes) with dogs and swine in *Khanda* 10, verse 7. The *Ramayana*, a significant holy book for Hindus, speaks of Lord Rama's rule (*Ram rajya*), during which only the three upper castes are allowed to do *tapasya* (penance and meditation) to attain divinity and dignity. Lord Rama, on learning that a *Shudra* had undertaken *tapasya*, killed him for being so presumptuous—one could only infer what might have happened had it been a Dalit/outcaste. Hindu scriptures such as the *Mahabharat* and the *Puranas* make reference to Dalits as monkeys (*banaras*), bears (*bhalukas*), and demons (*asuras*). It is the *Manu Smriti* (300 AD), popularly known as the Code of Manu (a sage), a collection of some eleven chapters with 2,419 verses (*slokas*), however, that instructs and commands into being a social order entrenched in the caste configuration. As the theological justification for caste/untouchability based on pollution–purity divides (e.g., nonsharing of well water or cooking utensils), caste endogamy, and constraints on freedom of movement/access (e.g., caste-segregated living arrangements and restrictions on entry to places of worship) (Gupta, 2000; Srinivas, 1996), these religious scriptures sought to ensure the karmic fate of Dalits for posterity in social hierarchies of caste—hierarchies that are notably conspicuous to date, with their attendant psycho-relational prescriptions and sociocultural and political-economic implications for the reproduction of Dalit marginalization, oppression, and exploitation.

The term Scheduled Caste (SC) was introduced by the British in the eighteenth century. Contemporary constitutional schedules list 1,116 SC groups for the purposes of constitutionally sanctioned amelioration, as SCs (a similar list of Scheduled Tribes, or STs, was developed for Adivasis/

original dwellers) account for seventeen percent of the Indian population (more than 170 million people). The postindependence Indian state outlawed untouchability as per Article 17 of the Constitution, while Article 46 commits the state to protect SCs as recognized marginalized groups, ensuring their economic and educational interests while protecting them from exploitation and social injustice. Despite these measures to address untouchability, according to the National Human Rights Commission (NHRC) (all India statistics): In 37.8 percent of government schools, Dalit children were compelled to sit separately while eating; in 33 percent of villages, public health workers refused to visit Dalit households; in 48.4 percent of the villages, Dalits were denied access to water sources (untouchability practice); in 27.6 percent of villages, Dalits were prevented from entering police stations; and in 23.5 percent of villages, mail was not delivered directly to Dalit dwellings (2005, p. 144). Attempts by certain Dalit organizations to have such practices addressed as human rights violations, racism and prejudice in international forums have been repeatedly dismissed by the Indian government on the grounds that "there is no need to apply external human rights mechanisms to what is essentially seen to be within the realm of cultural practice" (UN CERD, 2007, p. 3).

The political economy of caste in contemporary times also suggests that caste "continues to be deeply imbricated in the perseverance of poverty" (Guru & Chakravarty, 2005, p. 135), as the proportion of SCs classified as marginal landholders (owning about a third of a hectare or less/near landless) is over 73 percent today and the proportion of SC landless agricultural laborers is growing, while some 48 percent of rural Dalits live below the poverty line, compared to 32 percent of non-SC rural households. Of the estimated forty million people who are bonded laborers working in slave-like conditions to pay off a debt, fifteen million are children, the majority of whom are Dalits. Although comprising 16 percent of the Indian population, Dalits comprise fully three-quarters of the ranks of the poorest of the poor (Guru & Chakravarty, 2005). The mutually reinforcing effects of caste-class hierarchies sandwich Dalits in a structural trap defined by double discrimination.

Given these caste-class realities, in this chapter we consider Dalit social movement action undertaken by a Dalit-Adivasi solidarity movement, or Adivasi-Dalit Ekta Abhijan (ADEA), in South Orissa, representing more than 350,000 SC/ST people residing in more than 2,000 villages (when considering the widest zone of organized operation/influence) to address ADEA movement-identified issues pertaining to Dalit land and forest alienation, specifically as it relates to the historical process of Dalit political-economic marginalization and the reproduction of caste-class inequality in India. Based on long-term and recent movement-related developments concerning the politics of division instigated by state-corporate-caste interests, including (a) the role of divisive legislation and land and forest classification schemes; and (b) the alleged caste/religious communal violence[1] in the district of Kandhamal on the one hand, and the communal politics of resource usurpation (e.g., the recent neoliberal drive to promote Special Economic Zones (SEZs) in the Scheduled Areas or protected zones for ST/SCs) by the contemporary state-corporate-caste nexus (underlying causes) in South Orissa on the other, we elaborate on some movement-based reflections, knowledge, and learning (augmented with academic perspectives where pertinent) pertaining to this politics and its implications for Dalit claims to land and forests. We also consider some of what has been learned and achieved from ADEA counterhegemonic activism in relation to these hegemonic incursions, and the significance of movement-based knowledge for the movements and Dalit-Adivasi prospects in the region.

These insights into movement-defined issues and related activism are developed from the experiences and knowledge gained through direct action of the lead author of this chapter (and other ADEA movement leaders), who is convener of the ADEA and a research associate and founding member of the Dalit-Adivasi CRDS established in 2006. The coauthor is a research associate and founding member of CRDS as well, and has been working with the ADEA process since its germination in the early 1990s. The insights in this chapter are also informed by a PAR project (Kapoor, 2009a) (undertaken since 2006 by CRDS and the University of Alberta[2]) that has been simultaneously contributing to and examining Dalit-Adivasi movement learning pertaining to land, forest, and Dalit-Adivasi ways (own ways) in the region (Kapoor, 2009b; 2010; forthcoming).

ADEA Movement Praxis: Addressing Dalit Land and Forest Alienation

The germination of the ADEA movement process goes back to the late 1980s, when young Dalit-Adivasi activists moved from village to village using local art forms (puppetry, theater, folk songs, etc.) to broach Dalit-Adivasi sociopolitical issues with the people in an attempt to encourage organized action to address the same. The focus was and continues to be on the political-economic, sociocultural, and theological structuration of Dalit-Adivasi marginalization in Indian society and related avenues for change through politicization and persistent organized activism by a united and audible Dalit-Adivasi movement. Some fifteen years into the process, the ADEA has emerged as an organized Adivasi-Dalit political force in the Mohana block of Gajapati district, and a wider zone of influence and impact stretching into the peripheral districts, including Phulbani/Kandhamal. The support base for the movement comes from Kondh, Saora, and Paraja Adivasis, and Panos Dalits who belong to communities that together represent 62 percent of the population in the seven districts of South Orissa (also Scheduled Areas). At the state level, SC/ST groups account for over 40 percent of Orissa's population, according to the 2001 census, prompting some historians to refer to the state as *Udra Desa* or *Shudra Desa* (country of the Shudras and Ati-Shudras, or lowest of the castes).

The ADEA has succeeded in organizing and activating participating communities by paying close attention to several known and emergent interlocking movement issues of mutual significance to Dalits and Adivasis alike, while always giving space to and demonstrating solidarity with the participating social groups and communities to set independent agendas and take up actions specific to those agendas (e.g., Dalits and casteism/ untouchability and related marginalization, including land and forest concerns). Within this framework, the ADEA has prioritized and taken several actions to address land and forest alienation and dispossession of Dalits in particular. According to the latest National Sample Survey (NSS) of India (1999–2000) some 75 percent of the SC population of rural India are landless to near landless. The situation for Dalits in the ADEA region of South Orissa is, according to our own research, even bleaker.

The ADEA's political analysis of the situation of landlessness and restricted forest access for Dalits (and Adivasis) has been developed through a historical and contemporary analysis of the politics of upper caste/class state, and more recently, corporate collusions (conscious and/or unmediated), based on historical memory and years of direct action in the contemporary political scenario—activism aimed at disrupting and loosening the hegemonic grip of these sociopolitical formations. The marginalization of Dalits from land and forest is understood in terms of a politics of division and the myriad attempts to splinter Dalit-Adivasi/lower caste–class solidarity (*ektha*) to undermine the strength of potential opposition to industrial resource grabs and the perpetuation of caste-class rule implicated in the new state-corporate nexus since the neoliberal turn in the early 1990s. The politics of division ensures Dalit land and forest alienation, and is assessed by movement actors (ADEA movement knowledge/learning) in relation to some of the following possibilities (two examples of movement analyses—augmented by academic insights when pertinent—are shared here): (i) British colonial and continuing post-independence legislation and associated land/forest classification/ rights schemes; and (ii) caste/communal conflict, forced migrations, and the SEZ/industrial encroachment of rural land and forests.

Legislation, Land/Forest Classification Schemes and the Cultivation of Dalit-Adivasi Conflict and Division

The politics of division germinated during British colonial times as the Forest Acts of 1878 and 1927 and the Government of India Act of 1935 were devised by the colonial government to maximize the revenue yield aspects of land and forests and to supply the military, industrial, and commercial sectors. Tribal rebellions were commonplace, and the British appeased tribals by creating special enclaves for limited tribal self-rule (via the Agency Area Act), while ignoring the place of other forest-dwelling communities and simultaneously sowing the seeds for the detribalization of forests for colonial industrial purposes. By 1936, the British began the process of enumerating communities and splitting them along the lines of "caste" and "tribe" according to the Scheduled Tribe and Scheduled Caste Act, thereby separating Panos as a SC group from Kondh, Saora, and Paraja communities, who were recognized as ST—the dominant social groupings and communities in the ADEA region and in South Orissa. These categories formed the basis for a system of preferential reservations (e.g., education and government jobs), privileges, and differentiated status that favored one community over another. As stated by a Panos/Dalit ADEA leader in an interview, these categories

> began to instill a sense of otherness, difference, alienation, and estrangement. Groups that were major allies and inseparable friends through time were now in competition with each other under the schemes, a process that the post-independence government of India has expanded. Panos Dalits, once given SC status, were viewed with suspicion as competitors for forests and land, communication between communities became strained, as historical bonds of sharing, caring and cultural mixing gave way to jealousy, rivalry and estranged relations. (Interview notes, 2009)

The Agency Area Act, designed to appease militant tribals with limited autonomy and self-rule at the expense of other forest dwelling communities (non-STs) during British rule, was later reintroduced as the Scheduled Areas Act under the 73rd Amendment to the Indian Constitution and simply incorporated as the 5th and 6th Schedules of the Constitution. Under the purview of this act, STs in these Scheduled Areas were entitled to the natural resources (forest and land), and the entire chairmanship of the self-rule structure (from the village to the district level) was reserved for and vested in groups classified as STs. SC groups, on the other hand, lost their political, social, and economic rights due to this artificial constitutionally mandated classificatory scheme. This process was exacerbated in the early 1990s and at the turn of the century by the reopening of the Scheduled Areas Act for purposes of effective implementation of the act under the aegis of retired Indian administrative services officer B.D. Sharma (who filed a writ petition in the Supreme Court) and what became the *Bharat Jan Andolan*, or national-level NGO campaign to activate tribal self-rule. The campaign slogan of *Jungle, Jal, Jameen Hamara* ("Forest, water, and land is ours/for STs alone") created significant unrest and violence aimed at SCs in the forested interiors of the ADEA region and elsewhere as Dalits were forced off their meager homestead and personal cultivable pockets of land, leading to a mass migration of Panos Dalits from the forested interior to the industrial hubs as far away as Gujarat. Village burnings were commonplace, the flashpoint being the Mandrabaju incident in Mohana Block in the ADEA region, where an entire Panos Dalit village was permanently displaced (a few STs were shot dead by police) and sought shelter for almost two years in Mohana Tehsildar's office (magistrate-level revenue officer), a group whose eventual "disappearance" was left unexplained.

The promotion of state capitalism, especially since the New Economic Policy (1991) and neoliberalization of the Indian political economy (Menon & Nigam, 2007; Ray & Katzenstein, 2005), in the forest sector has perpetuated and exacerbated these trends. Dalits/SCs are disproportionately targeted (under the Land Acquisition Act, for instance), as these laws make it possible for the state to forcibly evict forest-dwelling communities for an undefined larger public purpose, as do the Forest Conservation Act (1980) and the Wild Life (Protection) Act (1972). Pimple and Sethi (2005) attest to the new contemporary reality that neoliberal land policies have made summary eviction of forest-dwelling communities a lot easier either through reservation, leasing of state land to industrialists, or the activation of a Wild Life (Protection) Act that has defined forest dwellers as the enemy, and through a system of demarcations of land and forests for national parks and sanctuaries that exclude forest-dwelling communities—what Nivedita Menon and Aditya Nigam (2007) refer to as "dispossession by law" (p. 69). Given the higher status accorded to ST groups under these various legislative constructions (and we have discussed the dire contradictory implications of the same for STs/Adivasis as well elsewhere (see Kapoor, 2009c; 2010)), the plight of SCs/Dalits is predictably worse when it comes to land and forests, given their secondary status under these legislative schemes and their caste-determined vulnerability. According to one Adivasi ADEA leader:

> The Dalits in South Orissa and elsewhere in India are deprived of their land and forest rights as caste society takes it for granted that Dalits are

by caste landless peoples and must remain so since they do not have a natural claim to land and forests. This is the Manuvadi mindset where Dalits are given the burden of tilling the master's land and as per Manu's law, never allowed to hold land or title to land. (Interview notes, 2009)

The latest legislative development, the 2006 Scheduled Tribes and Other Traditional Forest Dwellers (Recognition of Forest Rights) Act, or FRA, has been hailed by people's movements as a progressive development since it appears to recognize Adivasi/forest-dweller rights as a whole (some acknowledgment of "other forest dwellers"), including the right to their own agroforestry practices (e.g., swidden cultivation in the ADEA region) for local production and consumption. However, this is now being viewed in some quarters (e.g., by the ADEA movement leadership) as yet another "law and 'new welfare model' used by the State to retain its authority, power and supremacy over resources, alienate people from their land and way of life, and create and sustain capital markets" (Ramdas, 2009, p. 72). According to a government report in relation to the FRA:

> Now the tribals can cultivate their lands with dignity and without any fear. Tribals can plant rubber plants, mango, cashew nut, orange, lime, or palm oil as per local conditions. The state government would also develop lands in tribal areas and the tribals will be paid daily wages under NREGS program though they are working on their own land. (Cited in Ramdas, 2009, p. 69)

That is, the Ministry of Tribal Welfare and the Forest Department have interpreted the FRA provisions as a license to sanction export- and urban-market-oriented mono- cash-crop rubber, coffee, and fruit plantations in conjunction with the National Rural Employment Guarantee Scheme (NREGS), whereby tribals are reduced to being a source of cheap labor for these so-called tribal development schemes, and land and forests continue to be used for supplying export markets. Production for food by Adivasi (tribals), however, is being viewed by Forest Departments (for instance), as "encroachment" and is being met with considerable aggression to evict Adivasi/forest-dwelling communities from their homelands across the country, as is increasingly the case in South Orissa and in the ADEA region. According to an ADEA leader and Dalit commenting on the FRA:

> We are denied forest rights/claims because we are not perceived as being indigenous/Adivasi. For caste society, we are still a servant group meant

for doing labor and providing services, and only worthy of receiving handouts on special occasions. For our claims, we have to prove that we have been here seventy-five years or more based on papers that we have never been given in order to prove residency. The act has again differentiated between forest dwellers based on so-called ethnicity and created conflicts that can now be seen across South Orissa. (Interview notes, 2009)

Ironically, given the relatively more dire situation of Dalit landlessness and poverty, NREGS ends up being the main option for landless Dalits and migratory labor (the majority of whom are Dalits), and the ADEA has been torn between ensuring wage guarantees and access to NREGS for Dalits while recognizing how this might help to reproduce Dalit landlessness and forest alienation.

Land and forest classification schemes developed in relation to some of these acts have accentuated the divides between Dalits and Adivasis, as all claims are allowed only with respect to the shrinking category of *avadi* (claimable) and *avadi yogya anavadi* (claimable but not claimed) land. Other land and forests have been sectioned off for varied purposes, including reserve forests (*sangrakhit jungle*), revenue forest, waste/useless forest (*patita jungle*), hilltop forests (*pathuria jungle*), etc. The land settlements/claims process and outcome is antithetical to the traditions and customs of Dalit-Adivasi forest dwellers. It is based on individual private ownership, whereas people's customary practices are mainly collective, including a collective labor system, or *enthra*. These land and forest classification schemes make it difficult for Adivasi/ST claims, and almost impossible for Dalit/SC claims when coupled with acts (such as the Forest Rights Act) that privilege the former over the latter.

Since the government's revenue demarcation of land and forests, what belonged to all of us suddenly got divided into two *moujas*/areas of claim and people have started saying, this is mine and this is mine. Our neighbors are not allowing us to even set foot in their *mouja* and they are saying you should not cut our trees or bamboo for your use. And we are doing the same. This is not our way.... Our communities once had equal access to the land and forests, which today has been controlled using outside methods of the *sarkar* (government), and the *vyaparis* (business classes) and upper castes (*Brahmins*). (Adivasi elder, ADEA village meeting, 2007)

SEZs/Industrial Encroachments and the Role of Caste/Communal Conflict and Forced Migration in State-Corporate-Upper-Caste Land Grabs

In Orissa alone, "forest diversion" (a euphemism for industrial and development land grabs) in the post-neoliberal era has doubled between 1991 and 2004, while 26 percent of forestland cleared since 1980 has been after the introduction of the FRA (Wani & Kothari, 2008). This process of deforestation is being encouraged through the development of Special Economic Zones (SEZs)—duty- and tax-free enclaves to be treated as foreign territory for trade operations, duties, and tariffs. According to Environmental Protection Group Orissa, "These zones are going to be spread across thousands of hectares of common lands, forest areas, coastal areas and even on fertile agricultural lands" (cited in Menon & Nigam, 2007, p. 65). SEZs are also exempt from many labor and environmental laws, and even though there are some requirements for environmental clearance before work starts in an SEZ, the infamous case of Vedanta Alumina Ltd. and its bauxite mine on Dongria Kondh sacred lands in the Niyamgiri hills of Lanjigarh, Orissa, suggests a lack of seriousness about such stipulations. Similarly, when the tribals of Kalinganagar, Orissa, protested against acquisition of their land for a Tata (one of India's largest industrial firms) steel plant in January 2006, the state police (twelve armed platoons) opened fire on protestors and killed twenty people. Through the 1990s and increasingly today, private corporations are benefiting from state support (including armed support) for what ADEA members see as land grabs from Dalits, Adivasi, and peasants in general. As Menon and Nigam (2007) observe, however, another equally marked feature is "the growing resistance to such policies by dispossessed groups in different parts of the country" (p. 68). Kalinganagar, Lanjigarh, Kashipur, the anti-POSCO movement (the Pohang Steel Corporation of South Korea has signed large-scale mining- and steel-related MOUs, securing 600 million tons of Orissa's iron ore over thirty years), and Baliapal are but a few recent developments in Orissa that have been instructive for ADEA land and forest initiatives for Dalits and Adivasis alike. The recent outbreak of caste-communal violence in the Kandhamal district has taught movement participants about the intricacies of the politics of division for the reproduction of state-corporate-caste ruling relations and the continued land and forest marginalization, and impoverishment of Dalits in particular. ADEA leadership speculative analysis of these recent developments suggests that this nexus is losing patience with resistance to SEZ acquisitions and is subsequently resorting to or hiding behind the use of caste-communal violence to facilitate land grabs for SEZ developments and/or the privatization of land/forests for upper-/middle-caste business interests in the Scheduled Areas (ADEA region), specifically at the expense of Dalit Christians and some Adivasi Christians.[3]

> Company X owns a chain of shops/supermarkets where they sell vegetables and *haldi* (turmeric), ginger, mustard, and other organic food products to middle-class consumers in the towns. They have proposed the idea of having a supermarket in Berhampur, for which they plan on introducing large-scale contract farming and industrial plantations in Kandhamal, which is world-renowned for its turmeric, ginger, and mustard. People are now openly discussing that Corporation X is a major stakeholder for land in Kandhamal, and it is really important for us/ADEA to see whether there is a connection between the recent conflict in Kandhamal and these caste-business-state interests. In fact, our information tells us that the state government is already in the final stages of securing land/forest in this district to set up an SEZ for mass production/plantations of vegetables and spices. (ADEA Dalit leader, central meeting notes, 2009)

The local center for *Hindutva* (and Hindu right-wing groups known as the Sangh Parivar organizations) is the Orissa state-funded ashram of the late Swami Laxmananda Saraswati in Chakapad Block. The swami and the Sangh were instrumental in advocating for the voluntary and/or forced conversion of Dalit/Adivasi Christians to Hinduism. After the December 24, 2007, pre-Christmas denouncement of and attack on Dalit Christians, the subsequent assassination of the swami (publicly claimed by Maoist guerrillas, who had repeatedly warned the swami of such an eventuality should he continue to attack Dalits, but predictably blamed on Dalit Christians by the Sangh Parivar as an act of Christian retaliation), and the subsequent carnage of August 23, 2008, perpetrated by Sangh-affiliated groups against Dalit and some Adivasi Christians in Kandhamal district and surrounding areas, unleashed a reign of terror. According to CRDS estimates:

1. Some 40,000 Dalit Christians were displaced and forced to hide in the forests for months, many of them having abandoned their villages to this day (forced migration).
2. 5,000 or more dwellings were destroyed.
3. 600 or more churches and places of worship were demolished.
4. Some 25,000 people ended up in relief camps that were set up only months after the violence and after considerable pressure was applied to the BJD-BJP coalition state government at the time (the BJP is a

political party with public affiliation to the Sangh Parivar), which underplayed the impact of the calamity.
5. Forced conversions to Hinduism involved burning of the Bible and ritual purification through being compelled to drink a mixture of cow dung and water (Dalit Christians were tortured or killed if they refused).
6. Bodies recovered afterward showed signs of decapitation, torture, and burning, and victim statements shared with the ADEA confirmed the systematic use of gang rape and live burials. The Sangh-affiliated groups had a cadre of nearly 100,000 people in the region to carry out these attacks; state police and reinforcements seldom became involved in maintaining law and order, claiming that the interior was inaccessible, while stationing themselves in the main town areas, distant from the deadliest violence.

Now that close to 40,000 Dalits have left their villages, their meagre homestead land has been taken over by other caste groups who are willing to farm these lands (which are now described as 'vacant lots') as contract farmers to local traders and intermediate petty trader castes, who in turn will supply products to Corporation X until such time as SEZ status is confirmed and then they will willingly hand over these lots for money and the company will not have to put up with people's resistance to these land grabs for industrial agricultural plantations for export/city markets. Small landholding patterns get in the way of corporate expansion plans—the land has to be vacated by any means and voices that oppose also need to be removed before any commercial operation can commence.

A nexus has developed between the upper caste-run corporations, the trading communities/castes and Hindutva state-political interests who also represent industrialist interests like that of Corporation X and Y. They have used a section of the Kondh Adivasis who are close to Sangh organizations despite the fact that they are also *panchamas* like us (outcastes), to carry out the ethnic cleansing of Dalit Christians to enable such land takeovers from Dalits.

We have now seen how these so-called communal caste conflicts are being used systematically to protect and enhance the economic and political interests of the dominant castes and classes and to clean us out of home and livelihood. Some of our politicians are calling these activities development and saying it is time for us to change and to start to work for money alone so that we can buy our food from Corporation X's supermarkets and give up our land and food growing practices. (ADEA Dalit leader, central meeting notes, 2009)

ADEA Movement Activism: Addressing the Politics of Division and Dalit Land and Forest Alienation

The ADEA process in the region has grown from ten to fifteen villages in the early 1990s to some 2,000 villages (in terms of the widest zone of influence and organized presence) primarily because the emerging movement has been initiated and sustained by Dalits and Adivasis from the said regions who are well aware of the situation. The appeal of the movement also lies in its commitment to direct action aimed at the structural basis (caste-class domination exercised through the state-corporate nexus, for instance) of Dalit and Adivasi marginalization and exploitation, as opposed to the dominant NGO and charitable service response of programs, projects, accounts/budgets, contained/safe activism (referred to in NGO contexts as "empowerment"), and what some (see Aziz Choudry in this collection) refer to as an NGOization of protest and activism that channels long-term caste-class-related structural struggle into a techno-managerial set of poverty intervention schemes largely committed to working within (consensus-based) structural constraints of caste-class impositions responsible for Dalit land and forest alienation in the first place.

In response to the politics of division alluded to, the ADEA (as is the commitment symbolized by the name of the movement) has mobilized communities based on a popular message of "Dalit-Adivasi *ektha*"—a message that is carried through cultural mediums (e.g., songs and stories) that remind people of the unity of Dalit-Adivasi peoples through time and history, a unity that is vital in relation to addressing caste-class machinations of the state-corporate nexus, as in Kandhamal. The ADEA's commitment to *ektha* is made tangible through a network of village-based and regional organizational structures that work to educate, share ideas, and build on the *ektha* network in the region by pointing to the alternative—Kandhamal. Popular education and organizing (Kapoor & Prasant, 2002) form the bedrock of this process, and for every Kandhamal, there are many more examples of how similar conflicts have been preempted or nipped in the bud as ADEA villages have encircled areas with a show of strength; that is, people respond to a system of calling to form a wall, with country weapons in hand as a demonstration of confidence and strength, as they did in areas tested by the *Hindutva* onslaught of the Sangh in the recent past. Post-Kandhamal, more than 1,000 community volunteers have signed on as popular peace educators with the ADEA to help with critical analysis and the rebuilding of *ektha* in eighty *panchayats* in the affected and surrounding areas.

Dalit landlessness and forest alienation are recognized platform issues, and Dalit land "encroachments" have been prioritized in the ADEA's land-forest strategy. Dalits have mapped areas for Anavadi/other nonclaimable land (including "gray zones"—disputed classifications), and moved in en masse to occupy land now being farmed (e.g., growing traditional staples like finger millet or ragi) or used for the development of fruit orchards (local varieties)—a process that has placed more than 10,000 acres in "legal dispute" (the recent FRA is likely to enable the process of land titles). Many more "encroachments" have taken place and are awaiting legal redress, as the ADEA asserts itself over local forest, revenue, judicial, and law enforcement bodies. In fact, ADEA activists have developed special capacities and skills in identifying, demarcating, and then filing land applications, followed by effective advocacy thereafter to ensure titling. Since the land being occupied is adjacent or cumulative and proximate, even though legal titles are specific to individuals, the ADEA continues conversations around the importance of farming and using these lands in accordance with customary collective approaches and valuations, and has made some progress with this message in certain regions. Given the SEZ/industrial land-grab experiences in neighboring districts (including in Kandhamal), such as the ongoing struggle in Kashipur/Rayagada district, the ADEA has stepped up land claims and encroachments in a preemptive strike for Dalits and Adivasis, a process that is being reproduced at a quick pace given recent SEZ developments. The movement has also made huge strides in securing hutment area titles (*pattas*) for Dalits and Adivasis alike, which makes them safe from summary eviction as encroachers. Furthermore, since Dalits face an existential crisis of grave proportions, the ADEA has, despite expressed reservations that it is likely to perpetuate Dalit landlessness and provide industry with cheap and abundant labor, agitated for and filed several complaints through the legal process with regard to NREGS fraud and demanded compensation where Dalits have been paid less per workday.

Adopting a conflict-cum-cooperation approach with the state bureaucracy, the ADEA organized a mass demonstration in Mohana to protest the Kandhamal violence. State officials and the police were invited to attend—and they did—as the movement organization voiced its discontent (and exposed state complicity and apathy) with these agents and demanded redress for victims. With respect to the latter, the ADEA has worked together with the Human Rights Law Network in Delhi to bring culprits to justice, and is still helping victims file cases in police stations and maintaining pressure on the state to follow through on its promises of compensation. ADEA research and documentation, including testimonies and victim statements, have been used by the convener of the ADEA at the United Nations Human Rights Commission hearings on Kandhamal in Geneva, which was utilized as an international platform to point out the failure of the Indian state to act to prevent and stop the violence, contrary to the "India Shining" metaphor and image being used by caste-class elites to position Brand India abroad. In fact, several Dalit and human rights organizations and some NGOs have made use of these documents to publicize what happened in Kandhamal. Never underestimating the power of caste-class hegemonic forces and the state-corporate nexus, the ADEA recognizes the importance of reproducing the process of struggle while working to connect various struggles in South Orissa and across the state in order to address the scale of the politics of division and what it may mean for Dalits and Adivasis alike. Intermovement collaborations are at a nascent stage as some fifteen rural people's movements and movement organizations have been recently convened by the ADEA as the *Lok Adhikar Manch* (LAM) (see Kapoor, forthcoming) and with a great deal of political enthusiasm. LAM has already identified key issues for possible joint platforms and solidarity politics, while sharing analysis regarding progressive forces and opportunities, and is poised to become a counterhegemonic possibility in South Orissa, if not across the state.

The Significance of Knowledge Production and Learning in Dalit-Adivasi Social Movement Activism in South Orissa: Concluding Reflections

The knowledge gained and shared through direct action to address Dalit land and forest alienation (and other ADEA initiatives) has been instrumental in promoting mutual life learning among Dalit and Adivasi alike and continuous activism and definition of key issues that confront both communities in South Orissa. The learning around state-corporate-caste collusions and associated deployments of caste and religious conflict to divide and weaken Adivasi-Dalit opposition to land and forest grabs by these social groups is proving to be indispensable in uniting (*ektha*) communities, in strengthening the coalition of counterhegemonic forces opposing such attempts at undermining Constitutional safeguards, and for developing a response to caste-class-elite attempts to break or sidestep the rule of law. This is an ironic development, as appeals to the law are now being made in a bid to discipline the lawmakers and enforcers with some degree of temporary success.

In fact, such knowledge and learning by village communities is proving to be a catalyst for intermovement collaborations (translocal activism and the scaling up of oppositional movements through vehicles like LAM) that could prove to be a far more resilient and wider oppositional constituency than what isolated and localized struggles might be able to mount. This possibility is becoming increasingly clear from recent bids by the state and union governments to arrest the leaders and activists of people's movements (the anti-POSCO movement is a case in point) under the pretext of addressing what the prime minister has recently described as the greatest internal threat to India's security, the Maoist guerilla insurgencies in the forest belt. While it is easier for the state to declare violence against Maoists (given the historical standoff with these armed groups), the state stands to lose its *moral* legitimacy (Nandy, 2007), even in the minds of generally self-absorbed consumer castes and classes, if it declares an open war and unleashes its monopoly over the *legitimate use of force* against relatively peaceful people's movements of the poorest strata of Indian society, namely the Dalits and Adivasis residing in the Scheduled Areas. Purportedly then, LAM formations are a potentially greater threat to state legitimacy and growing state-corporate collusion in acts of development dispossession of Dalits/Adivasis.

In addition, knowledge production and learning in the ADEA has been stoking the revival of cultural and political traditions, and deliberately and consciously creating space for forums of political activism and analysis that were relatively muted or sporadically effective prior to ADEA attempts to mobilize people to address common concerns and issues through a process of organized, sustained, deliberate, and informed activism. The state-corporate-caste forces of Dalit-Adivasi exploitation are confronting and have been unwittingly constructing their nemesis in the form of these popular movement activisms (D'Souza, 2010; Jogdand & Michael, 2006; Menon & Nigam, 2007; Rajagopal, 2003; Ray & Katzenstein, 2005). We remain cautiously optimistic about the possibilities of influencing the state through direct action and consistent movement pressure to address the interests of Dalit-Adivasi forest and rural communities while holding the state accountable to the Constitution and state policies and laws. This is especially necessary when these legal mechanisms undermine caste and class as well as corporate control and privilege, and the state predictably develops legal amnesia or becomes activist on behalf of these controlling social castes/classes and institutions.

In the final analysis, what the ADEA hopes to accomplish goes beyond the specifics of the Dalit-Adivasi circumstance in South Orissa—it hopes to reeducate and inform the wider Indian public about Dalit-Adivasi realities and aspirations in order to address what Menon and Nigam (2007) see as ignorance (deliberate or otherwise) around rural and forest dislocations and dispossession and a tendency to construct the response in liberal terms, addressing the *individual* dislocated person:

> There is no understanding of communities as the subjects of dislocation or of ways of life that are destroyed. There is an abyss of incomprehension

on the part of the Indian elites toward rural and tribal communities. Ripping them out from lands that they have occupied for generations and transplanting them overnight in to an alien setting (which is the best they can expect) is understood as rehabilitation and liberation from their backward ways of life. (pp. 72–73)

Notes

1. At first glance, this was indeed the case as non-SC local/outsider caste groups launched targeted attacks against Dalit Christians in particular, and some Adivasi Christians (surface events/causes).
2. Dip Kapoor (University of Alberta) acknowledges the assistance of the Social Sciences and Humanities Research Council of Canada (SSHRC) for this research into "Learning in Adivasi (original dweller) social movements in India" through a Standard Research Grant (2006–2009).
3. See Report of the Indian People's Tribunal on Environment and Human Rights, 2006, on *Communalism in Orissa*, (edited by Angana Chatterji & Mihir Desai) for one perspective on Hindu right-wing activities in the state and its implications for Dalits/Dalit Christians in particular. Also see Tanika Sarkar (2005), *Problems of social power and the discourses of the Hindu right*, for an analysis of key actors and the rise of the Hindu right in electoral politics and beyond.

References

Chatterji, A., & M. Desai. (eds.) (2006). *Communalism in Orissa: Report of the Indian people's tribunal on environment and human rights.* Mumbai, India: IPT Press.

D'Souza, R. (February 8, 2010). The economics, politics and ethics of non-violence. *Sanhati.* Retrieved from http://sanhati.com/excerpted/2119.

Gupta, D. (2000). *Interrogating caste: Understanding hierarchy and difference in Indian society.* New Delhi: Penguin.

Guru, G., & A. Chakravarty. (2005). Who are the country's poor: Social movement politics and Dalit poverty. In *Social movements in India: Poverty, power and politics*, ed. R. Ray & M. Katzenstein, pp. 135–160. Lanham, MD: Rowman & Littlefield.

Jogdand, P., & S. Michael. (eds.) (2006). *Globalization and social movements: Struggle for a humane society.* New Delhi: Rawat.

Kapoor, D. (2009a). Participatory academic research (par) and People's Participatory Action Research (PAR): Research, politicization, and subaltern social movements in India. In *Education, participatory action research, and social change: International perspectives*, ed. D. Kapoor & S. Jordan, pp. 29–44. New York: Palgrave Macmillan.

———. (2009b). Globalization, dispossession and subaltern social movement (SSM) learning. In *Global perspectives on adult education*, ed. A. Abdi & D. Kapoor, pp. 100–133. New York: Palgrave Macmillan.

Kapoor, D. (2009c). *Adivasis* (original dwellers) "in the way of" state-corporate development": Development dispossession and learning in social action for land and forests in India. *McGill Journal of Education* 44(1), 55–78.

———. (2010). Learning from *Adivasi* (original dweller) political-ecological expositions of development: Claims on forest, land and space in India. In *Indigenous knowledge and learning in Asia/Pacific and Africa: Perspectives on development, education, and culture*, ed. D. Kapoor & E. Shizha, pp. 15–34. New York: Palgrave Macmillan.

———. (forthcoming). Subaltern social movement (SSM) post-mortems of development in contemporary India. In *Beyond development and globalization: Social movement and critical perspectives*, ed. D. Caouette & D. Kapoor, 39 pages. Ottawa, Canada: University of Ottawa Press.

Kapoor, D., & K. Prasant. (2002). Popular education and improved material and cultural prospects for Kondh Adivasis in India. *Adult Education and Development* 58(1), 223–229.

Menon, N., & Nigam, A. (2007). *Power and contestation: India since 1989.* London: Zed.

Nandy, A. (2007). *The romance of the state and the fate of dissent in the tropics.* New Delhi: Oxford University Press.

National Human Rights Commission. (2005). *Annual report, 2004–2005.* New Delhi: NHRC.

Pimple, M. & M. Sethi. (2005). Occupation of land in India: Experiences and challenges. In *Reclaiming land: The resurgence of rural movements in Africa, Asia and Latin America*, ed. S. Moyo & P. Years, pp. 235–256. London: Zed.

Rajagopal, B. (2003). *International law from below: Development, social movements and third world resistance.* Cambridge: Cambridge University Press.

Ramdas, S. (2009). Women, forest spaces and the law: Transgressing the boundaries. *Economic and Political Weekly* 64(44), 65–73.

Ray, R. & M. Katzenstein. (eds.) (2005). *Social movements in India: Poverty, power and politics.* Lanham, MD: Rowman & Littlefield.

Sarkar, T. (2005). Problems of social power and the discourses of the Hindu Right. In *Social movements in India: Poverty, power and politics*, ed. R. Ray & M. Katzenstein, pp. 62–78. Lanham, MD: Rowman & Littlefield.

Srinivas, M. (ed.) (1996). *Caste: Its twentieth century avatar.* New Delhi: Penguin.

UN CERD (2007). *Shadow report to the UN CERD 2007.* New Delhi: NCDHR.

Wani, M. & A. Kothari. (September 13, 2008). Globalization vs India's forests. *Economic and Political Weekly*, 19–22.

CHAPTER 13

Anjuman-e-Mazareen Punjab: Ownership or Death—The Struggle Continues

Azra Talat Sayeed and Wali Haider

Introduction

This chapter examines a land struggle initiated in 1999 by nearly a hundred thousand tenant farmers in more than twenty districts of Punjab, Pakistan's most populous province. The tenants have contested nearly 70,000 acres of agricultural land in some of the most fertile areas of Pakistan. Starting in two different districts, Khanewal and Okara, independently of each other, the struggle came together under the umbrella of the *Anjuman-e-Mazareen Punjab* (AMP—Punjab Tenants Association). The Khanewal struggle was carried out by tenants of the Punjab Seed Corporation (PSC), a semiautonomous Punjab government body established through a 1976 Act of Provincial Assembly for systematic seed production, procurement, processing, and marketing of major and minor crop seeds. Tenants of the military farm management began the Okara struggle over lands under military control that were used for operational purposes such as camping grounds, dairy farms, and the production of oats and hay. Collectively, this is one of the biggest land struggles to have occurred in Pakistan in recent years, and to date it has succeeded in stopping the military farm management from being able to wrest control back from the tenants united under the AMP.

It is often considered that the AMP started when the Okara military farm management wanted to impose a new contractual relationship with sharecroppers tilling agricultural land under its control. But the seeds of this movement were sown much earlier, during British colonial rule. During the British Raj, British missionaries settled peasants from various parts of Punjab were settled in Okara; they were granted tenant status under the Punjab Tenancy Act 1887 and asked to make uncultivated land productive. Sharecroppers were promised a sixty percent share of the produce for themselves. This land was given to the Royal British Army in 1913 under the Punjab Colonization Land Act 1912 for a period of twenty years. According to the AMP leadership, after the lease lapsed in 1933, no further extension was given to the British army. To the best of our knowledge, no documentation can be found in the Punjab revenue department to prove that military farmlands were legally controlled by either the British army after 1933, or the Pakistan army from 1947 (when Pakistan gained independence) to date. Although there are some discrepancies on this issue in different scholarly and popular literature, there are no major deviations in either. Saigol (2004) has reported that the lease expired in 1933 but that the agreement was extended for another five years, at the end of which the military ceased to have any legal claim to the land. The land actually falls under the jurisdiction of the Punjab provincial government, and the military farm management control over the military farmland in Okara is illegal, violating the 1912 Land Act. Being one of the strongest pillars of the country's elite, the Pakistan military can violate the law with complete impunity. Instead of land being given to the people, or at least to the Punjab government, the military farm management in Okara district, according to sources known to us, has allegedly taken control of 17,013 acres of land, spread over eighteen villages.

A similar situation exists in Khanewal district, where barren land had been made productive through people being settled by the British Raj. Land had been leased for ninety-nine years to Sir Robert William, who had promised that peasants would be given half of the land after a certain period. After independence, the Pakistani government lessened the period of the lease, which terminated in 1975. In 1977, Zulfikar Ali Bhutto's government gave 7,000 acres to the PSC. The only concession given to sharecroppers was the reduction of the proportion of the share of their harvest tendered to the government from fifty percent to forty percent. The PSC was able to overturn this decision by a court ruling in 1990, based on which peasants had to again give them fifty percent of their share.

Peasants working for the military farm management faced many abusive exploitative mechanisms. No doubt the long history of colonial and postcolonial abuse laid the groundwork for the AMP's land rights movement. Conversations about the movement with peasants invariably allude to the daily humiliations, injustices, and abject poverty they faced due to the exploitative tactics of the military farm management and the PSC. With the plaques over the tombs in their graveyard as witness to their history on this land, the peasants, although tilling it for nearly four generations, have been forced to give every breath of their labor to produce the maximum from the land that they had in the first place made cultivable. They have never been recognized for this service, nor given control over the fruit of their labor. Instead, they have been humiliated and treated like thieves, enduring conditions of oppression close to slavery. For instance, the military farm management system for computing the fifty percent share that the tenants had to hand over to them was not computed on the basis of each tenant's production. The harvest from the most fertile lands was considered as the average per acre production; fifty percent of this was what each tenant was forced to surrender, irrespective of their land's actual productivity. Nearly every season, farmers were forced into debt to be able to give the military what it claimed. Tenants also had to cut and load the harvest and deliver it to military farm warehouses. If one warehouse was already full, they were directed to go to the next. At most they were paid a meager sum for transportation, with no remuneration for harvesting and loading. Moreover, although the land had ample wood, tenants could not use it for their personal use, even for heating or cooking.

As in Okara, Khanewal tenants faced similar exploitation. According to them, their wheat harvest was taken to the PSC offices, and when they were given their share, the wheat had twenty to thirty percent dirt mixed in it. The PSC would tamper with the weights, decreasing the total amount received by the tenants. On paper it would state that they were given 110 kilograms of wheat. In reality it could be as low as eighty kilograms.

In this context of longstanding exploitation and oppression, in May 2000, the military farm management asked for a meeting with tenants, informing them that they wanted to change the tenancy system to a lease-based contract system. According to military officers at the meeting, the tenancy system was backward and corrupt, allowing both the officers and tenants collaborating with the management to fleece the common peasantry. The tenants initially spoke of a feeling of relief from the exploitative presence of the military farm management. But soon it became clear that this deal would eventually lead to them being evicted from land that they and their ancestors had tilled for many generations. Just after the farm management asked for the implementation of a contract system, it was announced that approximately 450 acres of land in four villages adjacent to the dairy farms (managed directly by the farm management) would not be considered under the contract system. Tenants working on these lands were not being giving the option of land lease, but would have been eventually evicted (Dogar, 2004).

This led to the smoldering mistrust felt by many generations of peasants against the management to ignite, kick-starting a movement known by its clarion call *"Maliki ya Maut,"* meaning "ownership or death." The AMP was created in June 2000, a month or so after the meeting with the military farm management. The *mazareen* movement, with a membership of nearly one million peasants from various military farms and the PSC, is present in many districts of Punjab. An estimated forty percent of its members are Christians. In a country where minority rights are poorly understood or protected, it is commendable that Christian and Muslim communities in the movement were able to forge bonds that held strong against many forms of vicious state abuse, or perhaps it is because of them.

The Anjuman-e-Mazareen Punjab: Rich Learning Ground for the Land Rights Movement

In order to understand the AMP's strategies for rebutting the military farm management and paramilitary forces, a number of interviews and focus groups were held with AMP leadership, activists, and peasants. We ensured that more or less the same level of information was requested from all participants. This was especially important, because after nearly a decade since the movement's inception, two different factions claim to represent AMP. Yet while carrying out interviews with both factions, there were negligible differences in their narratives recounting the historic events, strategies employed, and major challenges faced at the height of AMP resistance against the military farm management and the PSC. In addition, we draw on information from popular literature and more scholarly works (i.e., Saigol, 2004; Siddiqa, 2007). Most of the legal documents were gathered by the AMP itself, though systematic references can generally be found in mostly scholarly articles and some in popular literature. These include citations of government documents or correspondence, unearthed as part of the movement's work to resist the contract system imposed on them. For instance, a report titled *"Jag Utha Kissan,"* published by Pakistani nongovernmental organization (NGO) The Network, provided a scanned copy of a letter from the Board of Revenue in Punjab, to the Director General Remount, Veterinary, and Farms General Headquarters, Rawalpindi (Arif, undated). The letter, with the subject heading "Issue of Military Farms Okara," requests information on documents pertaining to the military farms.

It is clear from the recounting of milestones and events of the movement that the AMP embarked on a steep learning curve. At the very start, even before talk of a contract system, the military had forced tenants to acquiesce to decrease their share of the harvest from sixty percent to fifty percent.

However, final fear of eviction was the spur that led to determination not to let the military take control of their land. The movement's success can be attributed to several factors: the AMP's rapid organization, planning, and implementation of effective mobilization strategies and tactics; committed media coverage; engagement of the entire community, especially women; and research.

Research and Mobilization: Legal and Moral Victories

The movement drew strength from a long history of exploitation and abuse faced by the tenants. Its ability to translate this experience into a focused agenda of land ownership has been highly effective. The slogan *Maliki ya Maut* is now synonymous with AMP, articulating the most crucial demand and close affinity to the peasants' agenda, successfully gathering them under one umbrella. Even with the split within the AMP, there is no split in the demand.

The AMP leadership's clear-sightedness in trying to research the issue and find a legal basis for fighting for their rights was key to getting the movement mobilized so quickly. Although AMP leaders like Younus Iqbal played a vital role in painstakingly unearthing government documents to give them greater clarity on their rights as tenants, the involvement of journalists from the early stages of struggle helped the movement gain momentum. For instance, during a conversation on the issue of AMP, academic and human rights activist Rubina Saigol mentioned that early on, the leadership was only trying to find the contract that the British had made with tenants on the percentage of the produce that they would be entitled to.

The AMP leadership had researched the legal differences between tenancy and contract systems. This led to a clear understanding of their fate if the tenants accepted the contract system. The contract being imposed would allow military farm management absolute control over lands being tilled. According to Saigol (2004):

> [The] contract...would effectively change the status of the tenants from tenant to lessee.... The land would be leased out to the tenants for an initial period of seven years and the contract payment would be increased by 5 percent annually.... The conditions were extremely harsh and the tenant could be ejected from the land for minor infractions. For example,...the lessee could not chop or trim a single tree without the written permission of the Military Farm Management. Only the Farm authorities could decide which crops to sow, while the lessee would pay the abiana (water charges) as well as taxes. (p. 24)

Younus Iqbal remembers that the AMP had translated the demands being made on the tenants under the contract system. These included:

- Those who did not have agricultural land would not be given residential land.
- Contracted farmers could never demand ownership over land.
- Every year, there would be a ten percent increase in the lease.
- The administration could at any point terminate contractual obligations.
- Contract holders would not be allowed to participate in political activities.

Another major breakthrough had been unearthing the fact that the military farm management had no legal ownership over the military farms. Some of the top AMP leadership such as Younus Iqbal searched for government documents to clarify the land ownership issue. Because a large part of the AMP was Christian, they knew through oral history that the church was actively involved in settling the tenants during British rule. Unearthing government documents and being able to verify that the Punjab government technically owned the land gave them a clear mandate to challenge the military's claims over the Okara tenants.

Although AMP had been clear on their moral claims, the added legal weight gave them further incentives to build the movement. According to the AMP general secretary, Mehar Sattar, this information surfaced only after farm management insisted that tenants accept the contract system. AMP's approach in researching these issues and popularizing their findings was a major reason for the movement gaining strength.

Given that the land did not belong to the military, the management should have been paying revenue to the Punjab government. However, no papers could be found to show that the military farm had paid its dues. According to information shared with us by another AMP leader, David Zahid, a report was published by the Punjab Revenue Department in the *Roznama* (a daily newspaper), according to which the farm management was considered a defaulter of the Punjab government, liable to pay 0.8 trillion Pakistani rupees. AMP activists believe that the idea of a contract system was concocted to raise revenue to pay the Punjab government its dues. The finding that the land belonged to the Punjab government strengthened the "do or die" resolve of AMP not to surrender their land. As a resistance rebuttal strategy, in 2001 a revenue officer from the provincial government, instead of farm management, was given the government share. However, it was later discovered that these revenues were transferred into military accounts and considered part of the contract payment. Subsequently, no other share has ever been given to the government or the military farm management.

Media Solidarity

Journalists have played a particular role, and early on, the movement engaged the services of Anwar Javed Dogar, who worked for the Urdu newspaper *Daily Pakistan*. Dogar had already been covering rural issues before the movement took a distinctive shape. Upon his editor's request, for factfinding purposes, he had been holding open meetings with tenants opposed to the contract system in many parts of district Okara. Thus, he obtained firsthand information on the oppressive role of the military farm management against the *mazareen*. These were difficult times—in October 1999, General Pervaiz Musharraf became Pakistan's president through a coup d'etat. Dogar's commitment to the tenants' cause resulted in his extensive coverage of the Okara movement's issues, which rallied around the slogan "*jera wahway ohe khawae*" ("those who sow the seeds shall reap the harvest"). Dogar's affiliation with the movement was such that he was named the first president of AMP. He died in 2004, still a strongly committed AMP leader and worker.

According to Dogar, other newspapers and journalists played a progressive role in publicizing the AMP's struggle. As a result of his advocacy for the AMP, Sarwar Mujahid, reporting for *Nawa-e-Waqt*, was subjected to arrests and torture. Consistent beatings resulted in him losing an arm. As further punishment, the paramilitary forces brought many lawsuits against him, yet he has remained steadfast. Other newspapers, including leading English dailies, such as *The News* and *Dawn*, were instrumental in getting widespread media coverage.

Effective Internal Mobilization and Organizing

The AMP's rapid formation and the organization of tenants under different wings, including youth and women in all villages involved in the movement, were highly effective in mobilizing people. The movement's evolution was based on educating communities about the military farm management's long-term eviction plans. Tenants remember how they would jump on their bicycles to go from village to village to hold meetings and mobilize the village people. A group of ten would be formed to go from village to village to motivate and organize tenants to join the AMP. A women's wing sent separate teams for this purpose.

Although the movement was spearheaded by tenants from different religions, this difference was largely accepted and overlooked. In Pakistan, some conservative Muslims will not drink and eat with Christians; however, at AMP meetings, Christian tenants tell of Muslim tenants putting up water coolers and admonishing their people not to let the difference be felt by separating drinking and eating arrangements. This collective spirit in the AMP was very strong at the movement's outset, with the first and second tiers of leadership coming from both religions. However, now there is evidence of some division. It seems that the farm management has tried to exploit religious differences. Yet although today there are different factions, villagers believe that these differences are superficial and that if management starts another round of victimization, the movement will unite again. Some leaders believe that since both factions remain united in a call for land ownership, the split has not really weakened the movement. This view is repeated in many villages by tenants who believe that regardless of differences, they would never be divided in their determination to resist attempts to impose a contract system.

Culturally, poetry and songs are valued and listened to in community gatherings. A very strong part of the movement was internalizing and recounting the history of other movements and heroic deeds through poetry. For instance, the earlier peasant movement of Hashtnagar (in Pakistan's North-West Frontier Province during the 1970s) was often connected to the Okara movement. The AMP gave birth to poets such as Ulfat Dildar, known as *shair-e-mazareen* (the tenants' poet), whose passionate poetry detailed the atrocities of the military farm management. Dildar started writing and reciting poetry spontaneously only after joining the movement in 2002. Before this, he had not realized this skill. Even after nearly nine years, the poetry and poets are often remembered in evenings spent reciting the inspiring and evocative words written at the movement's height, many of which had been written when the paramilitary Rangers viciously attacked and caused several deaths and injuries.

AMP internal communication strategies for mobilization were very effective. When the police or Rangers approached a village, women would loudly bang their cooking vessels. This noise sent a warning to people nearby so that everybody could come to confront the state forces. As the movement progressed, the strategy was maintained using a different mechanism. All villages have mosques with loudspeakers. Sirens, generally called "hooters," were attached to these loudspeakers and used to warn and collect people against police and Rangers attacks.

The movement's history has seen many violent confrontations with the police and paramilitary forces, notably in October 2000 and in late 2002.

A common tactic of police and farm management was to file cases against many AMP leaders and workers. Apart from several deaths of movement activists, hundreds were jailed on false allegations of murder, treason, and armed attack against farm management. Human Rights Watch published a detailed report on the murders of AMP members as well as tortures inflicted by the police and military forces (Human Rights Watch, 2004). In order to overcome this daily repression, the movement created a system to collect funds. A fixed sum per acre holding had to be given to the leadership by each family associated with AMP, which was used to support families of those jailed.

Other farm management tactics included telling people employed in government service that they would be terminated if they did not leave the movement. The police and farm management cordoned off villages, which made people's commutes to the outside world difficult and resulted in many people losing their source of livelihood. The AMP leadership decided that instead of putting these people at daily risk, they should be asked not to go into town. These families were also supported by the collected AMP funds.

Some were forced to sign the contract in jail. But as a strategy, the leadership refused to honor those contracts, saying they had been signed under duress. One activist who spoke with us remembers that he had only signed the contract because four of his nephews were arrested with him. All were government servants threatened with termination from service, so he felt morally obliged to sign to save their livelihoods.

The Okara tenants are situated along a strategic highway, Grand Trunk (GT) Road, linking the provincial capital and Pakistan's second largest city, Lahore, to the country's south. Many AMP mobilizations blocked GT Road, causing immediate concern to the administration. These actions received swift media publicity, which was widely disseminated to all parts of Pakistan through word of mouth. AMP uses this tactic to this day in order to respond to injustices.

Women's Role in the Movement

It is uncommon for women to be in the frontline of leadership in a country as conservative as Pakistan. However, many women have played important roles in AMP leadership. These include Susan Bibi, Mariam Bibi, and Munawar Bibi, among others. There were many young women in the movement, including Aqeela Naz, Mumtaz Prem, and Rubina Albert, a fact that is even more commendable because young unmarried women are usually not even allowed to leave their homes, let alone be part of a highly risky movement against the military. Dogar detailed Mariam Bibi's courage in standing up to General Hasan Mehdi, an army general sent by the military farm management to hold a meeting at the Rangers' quarters in Okara. When he refused to let AMP leaders speak, Mariam Bibi challenged him:

> I don't know how big an officer you are. But I feel that you are a follower of those who had martyred the Prophet's grandson at Karbala. If you do not allow our leader to speak, we will give our lives, but will not agree to a contract system. The amount of money that you give your children as pocket money would be enough to feed our entire families. We are already dying of hunger, so if we have to die why should we not die with honor. (Dogar, 2004, p. 123)

According to an activist who spoke with us, people had agreed from the beginning that they could not win this fight without including women and children. Women of all ages had agreed that they would die but not give up their land. When the Rangers advanced, everybody would come out and stand on the streets. Generally, women and children formed the frontline, with men at the back. Women stated that they joined the movement spontaneously. The vicious attack on them and their families' livelihoods was the catalyst pushing them to part from their conventional roles and actively resist. Today they speak of the movement's difficult years, when their lives would revolve around the need to resist the farm management and the police from taking over their land. Conversations with them during the height of the movement showed their grim determination not to submit to pressure applied on them and their families. Susan Bibi, whose husband was in jail, stated clearly that she would not give up her and her family's rights over land. There are many stories of women's bravery. For instance, when Rangers once forced Susan Bibi into their van, other women surrounded it, forcing the Rangers to release her (Haider, 2002).

Women recall the heady days of constant strategizing, resisting, and movement building. Their activism occupied them fully, leaving them no time to care for their homes or children. They decided to stand in front of the men because if men were in the frontline they would be harshly beaten and arrested. Women would bring their *thappas*—laundry-washing clubs. Apart from their militant role in beating back the Rangers, they further mobilized women in the various military tenant farmers' villages. There is no literature on the Okara movement that does not detail women's involvement against the military dairy farms. According to Mumtaz and Mumtaz (undated):

> That women had become integral to the struggle was evident from women's participation in AMP protests. Women often lead these processions

> not only in Okara but also in Khanewal, Renala and other places. On 9 November 2001 they brought along their children, bedrolls and cooking utensils on to the main highway and camped on it. That day the Rangers fired at the protesters, killing and injuring several people. At the time of sieges women blocked all entrances to their villages and pitched their tents for weeks on end to make sure no one could enter and arrest the men.... Women were part of the hunger strikes and sit-ins, they moved from village to village, came to the cities, attended court hearings and addressed seminars and press conferences. Many were injured due to police action and firing and at least one died. (p. 7)

Women's active role is evident from the very first clash in October 2000, soon after AMP's creation. Farm management had sent a truck to pick up wood from a village, against the wishes of AMP leadership. On AMP's call, the entire community, with women leading, gathered where the wood was stored. They had brought *thappas* and forced the truck to leave empty. Later, the military farm management sent more armed paramilitaries and a violent confrontation ensued. A similar incident, as shared with us by one of the most prominent elected AMP women leaders of the Khanewal resistance movement, concerned the Pakistan Seed Corporation's decision to stop irrigation water and cut off phone lines and electricity. The resistance told the authorities that they would swarm the city with cattle if these services were not restored within forty-eight hours. This forced a meeting where, when offered water, while men drank, the women refused the same, stating that they would only drink once their children and cattle had water to drink. Due to such resistance, orders were passed to reopen irrigation water.

Women's active role can be further understood by other forms of victimization that they confronted. The farm management had coerced male tenants to force their wives not to participate in AMP. However, many women refused. Ultimately, six women were divorced by their husbands, but often these were under duress, as husbands and/or their family members were subjected to terrible physical and mental torture in jail.

Militant Activism or Civil Disobedience?

Although the *mazareen* movement is considered a militant one, where the tenants had dared to challenge the Pakistan military, not with arms and ammunition, but certainly with sticks and stones, the AMP believes itself to be a peaceful civil disobedience movement.

In considering this, we recall one serious confrontation in 2002, when Rangers shot dead Suleman Masih, a peasant. On August 24, 2002, people from all communities that were part of the movement, including women and children, were supposed to meet at a particular village to protest the paramilitary forces' constant pressure tactics. The Rangers and police had made it a practice to stop villagers to harass them about their identity with threats and verbal abuse. As villagers gathered to protest these violations and harassment, paramilitaries cordoned off the entire area, blocking their passage. Villagers were again told to stop their resistance to the contract system. The confrontation intensified and people were attacked with batons. Tenants retaliated by throwing stones. *Mazareen* from nearby villages came in great numbers, and police opened fire, leading to Suleman Masih's death, and many injuries.

In all of the confrontations between state forces and the *mazareen*, only the police ever used weapons. Since 2000, at least eleven *mazareen* have been killed by police and paramilitary action. Although women have sometimes used *thappas* against the police and Rangers, and men and children have thrown stones, the movement claims to be peaceful. This is an interesting claim since outsiders often consider the *mazareen* movement to be militant, mostly because of the women's use of *thappas*. In addition, their slogan of "ownership or death" implies a certain degree of militancy. However, members insist that this simply means that if the government does not agree to give them their land, then it should take their lives (Dogar, 2004). They are willing to face death rather than give up their land; the slogan does not imply taking other peoples' lives.

Social and Political Activists

The quick dissemination of the movement's issues strengthened it internally. It allowed others to get to know the issue and join hands. By 2001 to 2002, many social and political activists and civil society organizations, including NGOs, were involved. There was national and international coverage of the movement taking place in Okara, Khanewal, and Renala Khurd military farms and other villages. Although tenants drove the *mazareen* movement, people in the communities remember the support role of various NGOs and other social and political activists with gratitude and in solidarity. While some AMP leaders critique some activists and NGOs for using the movement for their own purposes, all seem to agree that coordination with NGOs helped achieve better coverage of the issue. Organizations helped to organize press conferences in Lahore and Islamabad. Additionally, interviews were arranged with BBC and Voice of America, which helped to get broader critical media publicity. NGOs in Lahore and Islamabad, where some of the mass mobilization took place, also helped in providing accommodation and food for AMP activists.

Many NGOs printed popular literature publicizing the movement far and wide. Women's NGOs made special efforts to distribute effective print material highlighting women's roles in the movement, with spectacular photographs portraying AMP women's militancy at the height of the conflict. NGO literature and interviews with the leadership and villagers highlights the legal aid and advice provided, especially by prominent human rights lawyer Asma Jehangir.

The Factions

The common belief is that the relatively recent division in the AMP is between Christians and Muslims. From discussions with the leadership and workers, however, it seems that the divide has emerged more from various stands which the leadership has taken over participation in referendums and elections. In 2002, after Musharraf seized power, he held a referendum trying to legitimize his leadership. Although there had been disagreement in the movement's leadership on voting for General Musharraf, the AMP had supported him because he had promised to give land ownership to the tenants if he won. However, after he "won," he failed to fulfill his promise. In 2008, the disagreement had become a chasm. By this time different groups in the AMP supported different political parties. Iqbal and his followers supported Rao Sikandar, a candidate from Muslim League (Q), a party created by Musharraf to weaken Nawaz Sharif's leadership in the Muslim League, which after the coup came to be known as Muslim League (N). David Zahid and his group supported Rana Rabbani from Pakistan People's Party (PPP). Mehar Sattar stood in elections as an independent candidate.

AMP has been unable to overcome its differences and reach a unanimous decision on its position on supporting mainstream political parties, which would suit its objective of getting land-ownership rights. The decision to support Sikandar could be considered the weakest—it was during Musharraf's regime that AMP faced the severest forms of state repression; to put trust in his leadership again was neither a good political decision nor a principled stand. Perhaps there could be some justification for seeking PPP support, but the PPP had not taken a stand to restore an independent judiciary, a position supported by a large people's movement in 2007–2008. A better political decision may have been to support Muslim League (N), as it would have meant supporting a party that before Musharraf's coup had deliberated giving the tenants land ownership. This latter party has always reigned supreme in Punjab and would have been a more pragmatic choice. However, Sattar's decision to stand as independent candidate also holds water, as there were fewer issues of holding another party accountable, and an independent candidate might give the AMP direct access to political decision making at the provincial government level. In 2009, Iqbal and Zahid had seemingly patched up their differences. However, the Iqbal-Sattar rift remains. Additional incidents have apparently split tenants along Muslim/Christian lines, a factor that could have been exploited by intelligence agencies that want the movement to wither away.

Conclusion

Yet in the final analysis, regardless of the leadership split, no faction has ever used any other name than that of AMP; there has never been a second call other than that of "*maliki ya maut*." The common peasant believes that those in the leadership can have their differences, but that the people will never allow for any other stand than claiming ownership over land. As shared with us by a leader of the AMP, people now know the benefits of keeping the fruits of their labor for themselves. Nearly ten years of keeping the entire share of production for themselves has taken away the pangs of hunger. There is some amount of prosperity. Much of this has translated into a belief that every peasant is a leader who will not give up her or his claim to the land.

The division in the AMP over which mainstream political party to align itself with stands in direct contrast to the otherwise clear-sightedness of the leadership. While the AMP has been very successful in its political struggle, it has been unsuccessful in strategizing its linkages with external mainstream political forces. This weakness highlights the tenants' negligible association or vision regarding a wider political agenda on peasant land-rights movements generally associated with challenging feudal or corporate control over land and other natural resources.

AMP has been fighting for a single-point agenda of gaining control over the land they have tilled. But the movement does not stem from a point of discontent with the broader mainstream political forces in Pakistan, which have historically maintained domination of an elite class of feudal lords, industrialists, and military in the country. By ignoring the despotic role of this elite structure in the dominant political parties, and trying to gain favor with one or the other, chasms have appeared in the AMP leadership. Perhaps this lack of a broader political agenda and failure to build alliances with forces that challenge the role of the dominant political system is a reason that the movement has not as yet faced the full brunt of military abuse and power. If the movement would move from the single demand for control over land tilled by tenants and start to challenge feudal control or the ever-increasing dominance of corporate agriculture in Pakistan, we suspect that the state's repressive forces and mechanisms would swiftly move to annihilate the movement.

References

Arif, M. (undated). *Jag utha kissan* [The farmer awakens]. Islamabad, Pakistan: Organization for Protection of Consumer Rights.

Dogar, J. A. (March 2004). *Okara kay gharib Mazareen ki jad-o-jehad nay akri hoee gardano ko khatam kar diya: haqook ki baziyabi ki manzil karib a lagi hae* [The struggle of the poor peasants of Okara has forced stiff necks to bow: The goal for realizing our rights is now closer]. *Qaumi Digest* (Lahore), 108–129.

Haider, W. (July-September 2002). Anjuman-e-Mazareen ki Haq-e-Malqiat kee jad-o-jehad [The struggle of the Anjuman-e-Mazareen struggle for their rights]. *Challenge* (Tel Aviv), 11.

Human Rights Watch (July 20, 2004). Soiled hands: Pakistan army's repression of the Punjab farmers' movement. Retrieved from http://www.hrw.org/en/reports/2004/07/20/soiled-hands-pakistan-army-s-repression-punjab-farmers-movement.

Mumtaz, K. & S. K. Mumtaz. (undated). *Women's participation in the Punjab peasant movement: From community rights to women's rights.* Retrieved from http://www.sidint.org/FILE_CONTENT/415-104.pdf.

Saigol, R. (July 2004). *Ownership or death: Women and tenant struggles in Pakistani Punjab.* Women in Security Conflict Management and Peace [Wiscomp] project on non- traditional security.

Siddiqa, A. (2007). *Military Inc.: Inside Pakistan's military economy.* Oxford: Oxford University Press.

CHAPTER 14

How Do You Say *Netuklimk* in English? Using Documentary Video to Capture Bear River First Nation's Learning through Action

Martha Stiegman and Sherry Pictou

Introduction

The Supreme Court of Canada's 1999 *Marshall* Decision recognized the treaty rights of the Mi'kmaq people to fish commercially, sparking a violent backlash from nonindigenous fishers across the region. The case was the result of generations of struggle for recognition of the eighteenth-century Peace and Friendship Treaties and the inherent rights they were meant to protect. But as Bear River First Nation has learned in the ten years since the Marshall Decision, treaty rights "recognition" in the maritime provinces, on Canada's Atlantic coast, is being enacted through a process of assimilating Indigenous Peoples into the neoliberal capitalist fishing industry. It is a process that has relied on the centuries-old divide and rule tactics—between First Nations, and between Indigenous and nonindigenous fishing communities—that have so fundamentally etched racism into Nova Scotia's social fabric. This process has solidified Department of Fisheries and Oceans Canada (DFO)'s control over fisheries management in the interests of furthering a neoliberal program of resource privatization and corporate concentration of ownership in the industry (Davis & Jentoft, 2001; Stiegman, 2009; Wiber & Kennedy, 2001).

Maori scholar Linda Smith (1999) describes imperialism as a "process of systemic fragmentation" (p. 28)—fragmentation of Indigenous Peoples from their lands, languages, and ways of relating to each other and the natural world—and as a project that has relied on the twin processes of colonialism and capitalism, a racist system of European control imposed in the interests of securing markets for resource exploitation. Smith's description resonates all too well in Nova Scotia, where the interconnected systemic racism and ecological crisis of today have firm roots in our colonial past. As Bear River First Nation's experience illustrates, colonialism is alive and well in Maritime Canada, although it takes a new form in the context of neoliberal globalization.

Globalization is eroding the political will and ability of nation-states to respond to local communities' needs. It is also creating new opportunities for alliance building. This has been the case in Southwest Nova Scotia, where the outrage of nonindigenous fishers, newly disenfranchised by neoliberal DFO policy that has seen massive deregulation of fisheries management and privatization of resources, has elicited a certain degree of empathy with their Indigenous neighbors. These were displaced by a much earlier colonial wave of enclosures that created the "public" resources the majority society holds so dear.

Many of us are learning that the key to resisting these twin threats is to realize a common cause between the struggles of First Nation and nonindigenous coastal communities. This solidarity is helping us build more effective resistance against the rampant resource exploitation and privatization that threatens the survival of all cultures. Winning the support of nonindigenous fishers has proven key for Bear River First Nation in its stand of resistance to government attempts to undermine treaty rights in the wake of the *Marshall* Decision. Building solidarity is not an easy task; through our experience, we are discovering that the work is as much cultural as it is political (Pictou & Bull, 2009).

In this chapter, we describe some of the cultural production that is helping build solidarity across communities that have been divided for centuries. We begin with the context of Mi'kmaq struggles for recognition of treaty rights and some of the ways this movement has intersected with nonindigenous fishers' resistance to neoliberal privatization over the past decade. We then describe the participatory methodology used for the production of the documentary film *In the Same Boat?* Finally, we look at the impact of this video process within Bear River First Nation, and we explore the wider communities of solidarity and resistance the documentary is helping to cultivate.

This chapter is itself a product and embodiment of these alliances. It is written collaboratively by Sherry Pictou, a grassroots community leader and former chief of Bear River First Nation, and Martha Stiegman, a nonindigenous video activist and doctoral student at Concordia University who grew up in the Nova Scotian settler community. Our relationship and the political analysis presented here have grown and deepened as a result of the collaborative video-based action-research project that we describe in the second half of this chapter.

L'setkuk

L'setkuk, or Bear River First Nation, is a tiny community of 150[1] at the headwaters of Bear River, which flows into the Bay of Fundy, famous for the highest tides in the world and which is a place of tremendous spiritual significance for the Mi'kmaq people. Traditionally, the way of life was migratory: People traveled throughout Kesputwick, the seventh traditional hunting and fishing district of the Mi'kmaq nation, in time with the seasons and cycles of life on which Mi'kmaq survival was so intricately dependent. Living in balance with all the creatures of Kesputwick was a responsibility given by the Creator.

In the Mi'kmaq language, L'setkuk means "water that cuts through" or "flowing along high rocks." This was a summer fishing camp where families gathered over the warm months after spending the winter dispersed, hunting across the territory. The name L'setkuk describes the trajectory of the river well, as it cuts a swath through the steep hills. It does not communicate the fact that the community was largely cut off from these fishing grounds and confined to a reserve in 1801, that this reserve is now a postage stamp of green in a sea of clear-cut logging, or that most of the fish and animals the community once relied on—bass, haddock, mackerel, salmon, moose, cod—are severally depleted or now extinct.

L'setkuk is also a stone's throw from Port Royal, where first contact with Europeans took place in 1604. The Mi'kmaq would be mostly displaced over the next 150 years, though around Bear River much of the traditional harvesting practices and lifestyle continued until the 1940s. Colonization is very old in this part of North America. The Covenant Chain of treaties that the Mi'kmaq and their allies negotiated with the British Crown stretches back to the 1600s (Grand Council of MicMacs, 1987), with the last of the Peace and Friendship treaties negotiated in 1761. These sacred compacts enshrined a vision of sharing the land as "two states sharing one crown" (Marshall et al., 1989, p. 82), with the Mi'kmaq adding an eighth point to the star symbolizing the seven traditional districts of the Mi'kmaq nation (Grand Council of Micmacs, 1987). As long as the sun shines and rivers flow, the Mi'kmaq would be free to maintain their way of life; in exchange they accepted the newcomers to Mi'kmaki. These promises were forgotten by the British no sooner than the ink had dried on the page. And so began the Mi'kmaq peoples' long-standing project of learning how to decipher the doublespeak of the Canadian government and to maintain Mi'kmaq values and practices while adapting the traditional way of life to non-Mi'kmaq economies, and of negotiating a balance between resisting colonial policies of extinguishment and assimilation while accommodating and integrating into nonindigenous society in a self-determined way.

Incredibly, though largely invisible to the majority society, the Mi'kmaq have survived despite more than 400 relentless years of colonization; despite the outlawing of traditional government under the Indian Act; despite the criminalization of the Mi'kmaq language and ceremonies until the early 1950s; despite the residential school at Shubencadie; despite Nova Scotia's attempts in the 1940s to centralize all the Mi'kmaq in the province on two reserves at Indian Brook and Eskasoni.

Court cases and police clashes provide a public record of Mi'kmaq resistance—from the trail of Grand Chief Syliboy, who was charged in 1928 with illegal hunting and referred to the 1752 Treaty to defend the Mi'kmaq's right to hunt and trap (*R. v. Syliboy*, 1928), to the 1973 and 1981 armed raids by Quebec Provincial Police and DFO wardens on Listiguj fishers defending their way of life (Obomsawin, 1984). But the headlines in the nonindigenous media fail to capture the spirit driving these events: the intention of Mi'kmaq people to live—as Kerry Prosper, an elder from Paq'tnkek First Nation would say—according to the laws that are rooted in the land of Mi'kmaki. For the Mi'kmaq, this vision is expressed through Netuklimk, a concept central to Mi'kmaq culture and worldview that "every living and non-living object was created equally, including humans. Everything in life is inter-connected. To sustain life in a respectful manner, lives must be lived responsibly and with consideration" (Prosper et al., 2004, p. 2).

The Marshall Case

This tradition of resistance is the context for the late Donald Marshall Jr.'s act of community-supported civil disobedience in 1993, when he went fishing for *k'at* (eel), a creature and food of tremendous ceremonial, medicinal, and spiritual significance (Prosper et al., 2004). Marshall was arrested for fishing without a license, out of season, and for selling his catch. His defense insisted, referring to clauses in the 1760–61 Peace and Friendship Treaties, on the Mi'kmaq's right to earn a living from the land (Coates, 2000; Wicken, 2002). The Supreme Court agreed, affirming the currency of the Peace and Friendship Treaties and the communal rights recognized within these for the Mi'kmaq, Maliseet, and Passamaquoddy to obtain a moderate livelihood through participation in the commercial fisheries. The ruling also recognized the Crown's prerogative to regulate such rights for the purposes of conservation, though the current regulations were deemed to be in violation of such rights in that they failed to explicitly acknowledge them (*R. v. Marshall*, 1999).

Marshall has been the political touchstone for events in Bear River First Nation over the past decade. As a reaffirmation of the currency and strength of Canada's treaty relationship with the Mi'kmaq, the decision is unparalleled. For Bear River community members, the decision was a deep affirmation of identity, of sacred attachment to the land, of Netuklimk and the way of life the treaties were negotiated to protect.

For nonindigenous fishers, the ruling was viewed as a threat and sparked a violent backlash across the Maritimes. The biggest headlines were from Esgenoôpetitj/Burnt Church, where shocking images of Royal Canadian Mounted Police (RCMP) officers beating Esgenoôpetitj fishers and DFO ocean cruisers ramming Mi'kmaq fishing boats made international news headlines for two summers (Coates, 2000; Obomsawin, 2002). But the backlash, which continues as a low-level conflict in many parts of the Maritimes, did not happen in a vacuum. It happened in the context of massive resource privatization and industrial overexploitation in the Atlantic commercial fisheries. In other words, the racism in Nova Scotian coastal communities is systemic: the legacy of colonial policies, and the evolution of capitalist relations and the current neoliberal restructuring in the fishing industry (Pictou & Bull, 2009).

Privatization and Resistance in NonIndigenous Fishing Community

DFO policy has long favored the development of a centralized, corporate-owned fleet capable of large-scale harvesting and processing for international trade, and has imposed industrial discipline on small-scale independent producers in order to integrate them into an ever-expanding and deepening capitalist market (Davis, 1991; Veltmeyer, 1990). This trend intensified in the 1980s after the influential 1982 Kirby Report (Canada, 1982), which laid out a neoliberal vision for restructuring fisheries, aiming to privatize rights to publicly owned marine resources, downsize the DFO, and deregulate management. This has been achieved primarily through the imposition of Individual Transferable Quotas (ITQs), a market-based approach to fisheries management intended to create market competition for control of quota, resulting in the survival of the most "efficient" and "competitive" fishers. As a result, Atlantic Canada has experienced a dramatic consolidation of corporate ownership in the fisheries and the near extinction of the family-owned businesses that characterized the small-boat fisheries for generations (Kerans & Kearney, 2006).

In 1995, when DFO threatened to impose ITQs on the small-boat cod fishery, coastal communities across Nova Scotia fought for and won the right to manage fishing quotas for their areas, and formed democratic organizations to coordinate community-based fisheries management at a local level. Around Digby, several organizations were created as part of this impetus, including the Bay of Fundy Marine Resource Center. The mobilization not only prevented a corporate takeover of the sector (though it should be noted that ground-fish populations have collapsed in the past five years because of larger privatization trends in the industry). Through that experience nonindigenous fishers also developed an intense distrust of DFO and a critical analysis of its privatization agenda, which would help lay the ground for dialogue with their Mi'kmaq neighbors after *Marshall*.

Although Bear River First Nation is only a twenty-minute drive from Digby, where thousands had taken to the streets protesting ITQs, this was a history of struggle completely unknown in the Mi'kmaq community, which illustrates just how effective de facto racial segregation is in Nova Scotia. Though the outcome of *Marshall* in most of the region further entrenched these divisions, around Bear River First Nation it helped foster change.

* * *

> I held that feather in my hand and realized, because I'm Acadian, you know, and my people survived deportation by the British (in 1775) because we were hidden in the forest by the Mi'kmaq. So I held that feather and thought about that history and what my grandparents would want me to do now.
>
> —Nonindigenous fishing leader, interview, 2003

While the media focused on the clashes at Esgenoôpetitj/Burnt Church, in Yarmouth, an hour's drive from Bear River, a potentially more explosive conflict was brewing. The entire Southwest Nova fleet, roughly 700 boats, blockaded the harbor in a show of force to keep Mi'kmaq fishers off the water. Politicians warned nonindigenous fishers that the Mi'kmaq would destroy their livelihoods; reporters stoked the flames by refusing to cover any constructive dialogue. Violence seemed imminent and tension mounted daily. A secret behind-the-scenes meeting was arranged between nonindigenous fishing leaders and the chiefs from the two First Nations in the area in an attempt to defuse the crisis. Frank Meuse Jr., former chief of Bear River, walked into that meeting with an eagle feather and asked that the meeting be conducted as a talking circle,[2] that everyone put aside the issues of the moment and speak from the heart about what their grandparents would tell them to say.

That sharing circle not only averted a violent crisis, but was also a deeply transformative experience for all involved—an emotional moment of empathy and of deep cross-cultural learning that lay foundations for further dialogue and eventual collaborative actions. It is an incredibly powerful story that has become a teaching tool in its own right, and several reflections stand out. One has to do with the culture of meetings and the importance of sharing not just our political analysis of the issues we face, but also our humanity—our hopes, fears, stories, and cultures. Another relates to the importance of overcoming the systemic racism that has divided us. The conflict resolution achieved in a hotel conference room outside Yarmouth that day happened without interference from government bureaucrats, lawyers, or negotiators. It happened face to face between local chiefs and community leaders from grassroots fishers' associations created by popular mobilizations. These democratic organizations, along with the safe, neutral location provided by the Marine Resource Center, provided a space for harvester-to-harvester dialogue outside the pressure of DFO consultations or other official political negotiations. This has proven critical to circumventing government and industry's divide-and-conquer approach in the wake of *Marshall* (Stiegman, 2009). The relationships established between Bear River First Nation and neighboring nonindigenous fishers through that initial conflict mediation have evolved, and we have since joined forces to oppose other forms of privatization in the area, including a proposed mega–rock quarry and the recent privatization of fourteen local beaches, which is displacing clam harvesters (Wiber & Bull, 2009). It has happened very differently in other parts of the Maritimes.

Government's Response to Marshall: Divide and Conquer

The government response to *Marshall* was twofold. Over the long term, the parameters of a treaty-based fishery are being established through formal negotiations involving First Nations and the federal and provincial governments as part of a larger process to implement the historic Peace and Friendship Treaties in a modern-day context. In Nova Scotia this is being carried out through the Mi'kmaq Rights Initiative (MRI). In the short term, DFO negotiated interim access agreements on a band-by-band basis,[3] offering funds to access communal commercial licenses, vessels, gear, and training. In exchange, communities agreed to shelve (Milley & Charles, 2001) their right to manage their fisheries for the duration of the agreements and to fish by DFO regulations. To date, thirty-two of the thirty-four First Nations affected by the Marshall ruling have signed interim access agreements; Bear River First Nation is one of two communities that refuse.

Bear River's reasons for doing so are many and extend far beyond a simple discomfort with accepting federal jurisdiction over harvesting activities and treaty rights. The capacity and legitimacy of DFO to act as environmental steward is questionable given the collapse of cod stocks and the department's neoliberal program of fisheries privatization—both of which are decimating coastal communities across the region (Kerans & Kearney, 2006). While fishing agreements are supposedly without prejudice to the exercise of treaty rights, Bear River First Nation's concern was that these agreements would lay the foundations for the aboriginal fishery being negotiated within the MRI, and retroactively be considered consultation and compensation regarding the infringement of treaty rights within those negotiations. Most important, Bear River finds it impossible to express its spiritual and cultural values through a fisheries management regime predicated on resource privatization, individual property rights, and hostility to the contributions of Mi'kmaq traditional knowledge.

Critics of DFO's privatization agenda were hoping that *Marshall*, with its affirmation of treaty rights and the creation of a distinctive Mi'kmaq fishery, could act as a crack in the dam of DFO's fisheries privatization agenda. The hope was that a progressive coalition of Mi'kmaq and nonindigenous fishers advocating community-based fisheries management could challenge the status quo, and that the management regime would be forced to incorporate a diversity of local regulatory schemes within a broad strategy for conservation (McIntosh & Kearney, 2002). What has actually transpired is quite the opposite.

DFO made room for First Nations entrants into the fishing industry by buying licenses from commercial fishers and then making these available to First Nations. These two processes happened separately, behind closed doors. In many instances this further entrenched divisions between First Nations and nonindigenous fishing organizations. It also isolated First Nations from the critical analysis of fisheries privatization that nonindigenous communities had developed through their struggles with DFO. The department's inflexible approach, the rushed pace of negotiations, and First Nations' lack of knowledge about the commercial fishing industry gave DFO effective control over the negotiation agenda. As a result, First Nations in Nova Scotia are given little more than local control over the implementation of DFO policy in the commercial aboriginal fishery.

In justifying its actions, DFO has referred to *R. v. Marshall* (1999), which acknowledges the department's prerogative to regulate aboriginal rights in the interest of conservation. From Bear River's perspective and that of other commentators, however, the department's primary motivation has been to maintain control in order to further a program of fisheries privatization (Davis & Jentoft, 2001; Stiegman, 2009; Wiber & Kennedy, 2001).

We have come to a juncture in history where the very resources that sustain the circle of life are in danger of collapse. While the lawyers and bureaucrats negotiate treaty interpretation, Bear River First Nation is engaged in a grassroots process of cross-cultural relationship and alliance building. With the colonial history and legacy of racism that plagues us to this day, why follow such a strategy? On a study tour through British Columbia coastal communities in 2002, a Nuu-chah-nulth Tribal Council elder put this into perspective by reminding Bear River participants that our responsibility is to take care of *all* of life in our traditional territories—including nonindigenous people. All of life is integral to Indigenous Peoples; therefore, all of life must find a way to live in balance.

In the Same Boat?

It's called displacement; we know all about that. I mean look at the handliners (hook-and-line fishers)—a whole way of life ended. We can relate to that—we have 500 years of relating to that.
—Bear River First Nation harvester at a community film screening (2007)

This was the context for the production of a documentary film: intercultural dialogue established in relation to fishing; community leaders having had transformative learning experiences but persisting racism and ignorance within the wider nonindigenous community; common ground established in relation to resisting the ravages of neoliberal privatization and joint political actions undertaken, but limited understanding of the treaties, let alone the inherent rights of the Mi'kmaq as Indigenous Peoples. Within Bear River First Nation, there were varying degrees of ownership of the community's political stand, a political position increasingly marginalized and invisible. Our hope was that *In the Same Boat?* would deepen these emerging dialogues, both within and across communities. In this section we describe the process of making the film, the impact of the collaborative methodology used in terms of transformative learning both within Bear River First Nation, and the widening circle of solidarity the film is helping to cultivate.

Participatory video is a method of video production in which the filmmaker engages the "subjects" in the project of deciding what story they want to tell, how, and to whom. The filmmaker sheds the role of auteur and becomes a trainer and social animator in order to make video *with* not *about* people marginalized by mainstream media. The process can be transformative and empowering as it engages participants to reflect on, analyze, and present their experience as a form of political action—an exercise that entails questioning assumptions about power relations, claiming a voice in public discourse, gaining skills and confidence, and building networks of support that can help lead to other forms of action (Rodriguez, 2001).

Within this practice, there is a wide spectrum of films whose form, point of view, audience, and aesthetic are adapted both to their cultural and political context, and to the goals and priorities of those involved on both sides of the camera. In some cases training and mentorship are integral, with the aesthetic of the final product secondary to the impact of that process within a community organizing initiative. Other films have a very focused message in relation to a specific campaign goal (for example, see WITNESS' model of video advocacy; http://www.witness.org).

In other productions, as for *In the Same Boat?*, the filmmaker maintains the role of director, assuming aesthetic and structural decisions for the film, but there is a shared authorship with film participants. This involves a collaborative process to arrive at the right questions to ask, and recognition on the part of the filmmaker of participants' agency to decide what parts of their lives they want to share—explicitly (what they want filmed) and implicitly (how they "perform" their lives for the camera). This is a slow process fueled by trust. In the case of *In the Same Boat?*, it took us roughly two years.

Project Design and Process

What became a two-part documentary began as two parallel short films. *The End of the Line* chronicles the struggle of nonindigenous hook-and-line fishers against DFO's privatization agenda. *In Defense of Our Treaties* explores the vision guiding Bear River's political stand and work developing a fishery grounded in Mi'kmaq values and knowledge. Our hope was that the process of collaborative film production would deepen local discussion about the grounds for solidarity established between both groups without glossing over their very real differences. We wanted to frame the question of common ground established around fisheries privatization in a way that would give voice to the Mi'kmaq perspective, which is still so misunderstood in the nonindigenous community.

Before shooting began, Martha Stiegman spent a month in Bear River First Nation working with harvesters and community leaders to establish the general content for a potential film. We agreed to begin an open-ended process: People could withdraw from the project at any time. They also had veto power over any material they did not want to appear in a final product. We spent a month shooting at the end of that summer; Martha returned in the spring to screen initial edited sequences.

That first focus group brought together harvesters and community leaders to view the initial footage, offer feedback, discuss the issues it raised, and decide if we should move forward and turn these initial sequences into a film. If, as they say, a documentary film is really a record of the relationship between a filmmaker and the people on camera, that exchange was the moment our relationship began. The visual, immediate, and accessible quality of video opened a two-way communication that allowed people to participate in the process of their own representation. It also gave participants the chance to give informed consent to move forward with the project and turn these initial sequences into a film. One person remarked, "Oh, I see where you're coming from now. I guess I won't have to hold my tongue around you anymore!" Through that group discussion and subsequent one-on-one conversations, people gave crucial feedback on the point of view of the film, on material that should be cut, and on missing elements of the story. Together we identified changes and a to-do list for a subsequent round of shooting later that summer. We worked together to establish what Mi'kmaq songs to include and which locations to shoot. Martha returned again that winter to present a full-length rough cut of *In Defense of Our Treaties*. There was another series of vetting sessions—first with the people in the film, then with the political leadership of the community. Meanwhile, a parallel process was being carried out with fishers and community leaders in the neighboring nonindigenous fishing community for *The End of the Line*.

Once both films had met with their respective community's approval, we organized a joint screening at the Bear River First Nation to see if and how these two stories worked together. The screening was open to everyone in Bear River, and invitations were sent to half a dozen nonindigenous fishing leaders in the area. Forty people spent six hours watching the films, discussing the issues raised, and debating who else should see them and why. There was a unanimous sense that the two films were really two sides of a single, larger story, and so those two parallel shorts became the two-part film *In the Same Boat?* We now turn to the impact of the collaborative film production process within Bear River First Nation before considering how *In the Same Boat?* has helped strengthen the grounds for solidarity.

In Defense of Our Treaties

The collaborative production process for *In Defense of Our Treaties* opened a series of conversations: individual in-depth interviews, the exercise of choosing how to represent oneself on camera, focus group meetings to respond to that exercise of representation and to discuss the themes explored, as well as larger community screenings. These discussions created a unique reflective space for analysis and dialogue about the community fishery BRFN is working toward—outside of the structure of band meetings or the analytic confines and pressures of responding to political crises or DFO demands. This provided a space to recognize not just the hard work of harvesters, but also the important role they play as traditional knowledge keepers. Weaving together each person's thoughts and experiences in a single coherent narrative strengthened a feeling of unity and purpose within the community. After viewing a rough version of the film, one participant commented, "I don't think we realized just how much on the same page we all were!"

Cumulatively, this legitimized the community's experience, helping to turn a perceived negative into a positive. Fishing had been thought of as something the community was not doing: BRFN was *not* signing an agreement, *not* developing a commercial fishery, *not* getting out on the water and making money like other bands. This was reframed: BRFN *is* making a principled stand, *is* articulating a unique and important vision, *is* adapting Mi'kmaq knowledge and values in a modern-day context. As one person commented during a community screening, "For us who were in the video, it's sort of a reflection—you don't think you're doing anything. You get so demotivated and tired of talking about the fisheries... but then you look at this video, and you see you're actually doing something—you're taking such a stand!"

It is impossible to quantify the impact of such a process, but our feeling is that the series of reflective spaces that *In Defense* opened up has deepened the community's understanding of the political stakes motivating the stand and strengthened harvesters' commitment to BRFN's project of building a unique, community-based fishery grounded in Mi'kmaq values and cultural practices. Proof of this lies in harvesters consistently prioritizing the development of low-impact wood lot management projects and fish habitat restoration work over efforts to engage with DFO to develop commercial fishing.

For BRFN, the video became an eagle feather or talking stick, giving the community a voice in a cultural and political environment that has turned a deaf ear to its perspective. It is a witness to the stand to defend our way of life, much as the treaties were for our ancestors who signed them in the eighteenth century. Two hundred years from now, there will be a record of the stand that is being taken here and of the vision guiding it. It is a deep affirmation of Bear River's struggle to maintain and live out the remnants of the ancestral, traditional Mi'kmaq knowledge held so dear—knowledge we believe is critically important for the cultural and ecological survival of all the peoples who live in our traditional territory.

Much has been written about the difficulty of translating traditional ecological knowledge (TEK) into a Western framework. From a holistic, Indigenous perspective, "knowledge" is not separate from culture, ceremony, or story; to label this as "traditional" freezes and reifies authentic Indigenous culture as something existing in the past that cannot evolve, incorporate elements from other traditions, and make relevant contributions to the present (Nadasdy, 2003). This has been an ongoing challenge for BRFN, whether it involves explaining Bear River's vision to the university researchers with whom we collaborate; negotiating with DFO, to whom the community tries to explain its position or find ways to fit its activities into the arbitrary categories of "food fishing," "commercial fishing," or "habitat restoration"; or within the community as it wrestles with internal colonization and the difficulty of articulating its values in English as it reclaims the Mi'kmaq language, which in BRFN has been all but lost. Yet something about the visual, narrative medium of video has allowed us to capture a glimpse of that vision and to document and share BRFN's Indigenous experience with outside audiences.

Deepening Solidarity through In the Same Boat?

> We're telling a story—it's not about the losses, it's about what was done. It's about our grandchildren being able to say, "Something happened here and our grandparents did something." It's not just about who won, its about what we did—this movie is for those who will come later.
>
> —Nonindigenous fishing leader at a community film screening (2007)

We now turn our discussion to the process of dialogue that *In the Same Boat?* is contributing to, beginning with that initial screening at the BRFN band hall that brought together community members from BRFN and nonindigenous fishing organizations in the area. The sense in the room after watching the videos was of overwhelming identification. People recognized that a neoliberal globalization agenda playing out in the region is privatizing the land and the waters, displacing small-scale fishers, decimating the natural resources both communities depend on, and threatening both cultures' ability to pass knowledge and tradition down from one generation to the next. This nonindigenous fisher's comments sums up the tone of the discussion:

> Watching these two films, it becomes apparent our common enemy is our government and DFO. You in Bear River have been fighting with the government, [nonindigenous fishers have], and we're no further ahead than thirty years ago, and why? DFO wants to keep us separated to give the fisheries to a few companies. Our battles are the same!

A response from a Bear River community leader highlighted the importance of cultivating solidarity, not just around fishing, but also Indigenous rights:

> We have been trying to explain what our treaty, and our title, and our rights are—and trying to get your support, saying that our community values are the same, if not identical to yours.... If our rights and our title are recognized, that gives us leverage to sit at that table with government, so that the next time they revise the Fisheries Act there's going to be First Nations sitting there, bringing our values. We lost that with Marshall, so we have to go back and build our case again and go back to the courts—but we need your help.

Community spokespeople have since toured *In the Same Boat?* through Mi'kmaq and nonindigenous communities across the province. While Bear River–area screenings presented an opportunity to deepen discussion of the issues related to fishing struggles, regional audiences—even those uninvolved in the fisheries—also expressed a sense of identification with the theme of being "in the same boat." People expressed a general sense that the neoliberal policy imposed on fisheries is the same agenda playing out in health care, education, and government generally. This current wave of neoliberal enclosures is privatizing the public resources, goods, and services the majority society in Canada holds dear. The sense of disenfranchisement and loss of sovereignty that Canadians now face echoes the colonial reality the Mi'kmaq have been dealing with for centuries as a result of the colonial wave of enclosures that swept across Mi'kmaki five hundred years ago.

The series of conversations that *In the Same Boat?* opened, both through the participatory production methodology and via community screenings, has allowed us to build from this sense of empathy, deepen our political analysis of how the current wave of neoliberal enclosures affects First Nations and nonindigenous communities, and make explicit the tacit lessons that we have learned through our decade of shared work. Film production has allowed us to approach these questions through the lens of culture and experience, and in so doing, to deepen the empathy, political analysis, and solidarity that ground our common political work. That work continues: Locally, this is being done through our current struggle against the privatization of beaches and the displacement of clam harvesters in Kesputwick (Wiber & Bull, 2009), and through the friendships that have evolved from our decade of collaboration. Nationally and internationally, we participate in learning circles with Indigenous and nonindigenous harvesters and researchers (both within and outside the academy), where through bimonthly conference calls and annual gatherings we are comparing struggles against privatization, colonial policies, and the intersections between these forces (see http://clcn.seedwiki.com). Finally, Bear River First Nation's participation in global networks like the World Forum of Fisher Peoples and Via Campesina (see http://www.viacampesina.org) allows us to link our struggle with peasant and Indigenous Peoples' movements fighting for recognition of small-scale traditional fishing and land rights internationally.

Notes

1. There are roughly 300 registered band members of Bear River First Nation, approximately half of whom live on the reserve.
2. The use of talking circles emerged from an indigenous democratic practice of uninterrupted speaking in council gatherings. Many Indigenous Peoples also use the talking circle for sharing and healing. Taking turns, the speaker holds a "token" or sacred object such as an eagle feather, which is passed on to the next participant. While the object denotes the speaker, many objects are considered sacred and thus provide strength to speak from the heart.
3. Band refers to the First Nation-level unit of government imposed on Indigenous Peoples by the Canadian government's colonial Indian Act.

References

"*R. v. Marshall* [No.1]." [1999] 3S.C.R. 456
"*R. v. Syliboy* [1928]" 50 C.C.C. 389
Canada Department of Fisheries and Oceans [DFO] (1982). *Task force on Atlantic fisheries. Navigating troubled waters: A new policy for the Atlantic fisheries: Highlights and recommendations*. Ottawa: DFO.
Coates, K. (2000). *The Marshall decision and native rights*. Montreal & Kingston: McGill-Queen's University Press.
Davis, A. (1991). Insidious rationalities: The institutionalisation of small boat fishing and the rise of the rapacious fisher. Retrieved from http://www.stfx.ca/research/gbayesp/insidious_report.htm.
Davis, A., & S. Jentoft. (2001). The challenge and the promise of indigenous peoples' fishing rights—from dependency to agency. *Marine Policy* 25(3), 223–237.

Grand Council of Micmacs, Union of Nova Scotia Indians, Native Council of Nova Scotia (1987). *The Mi'kmaq treaty handbook.* Sydney & Truro: Native Communications Society of Nova Scotia.

Kerans, P., & J. Kearney. (2006). *Turning the world right side up: Science, community, and democracy.* Halifax: Fernwood.

Marshall, D. Sr., A. Denny, & S. Marshall. (1989). The covenant chain [of the Mi'kmaq]. In *Drumbeat: Anger and renewal in Indian country*, ed. B. Richardson, pp. 71–104. Toronto: Summerhill Press.

McIntosh, P. & J. Kearney. (2002). *Enhancing natural resources and livelihoods globally through community-based resource management.* Proceedings from the Learning and Innovations Institute, November 6–9.

Milley, C., & A. Charles. (2001). *Mi'kmaq fisheries in Atlantic Canada: Traditions, legal decisions and community management.* Paper presented at the People and the Sea: Maritime Research in the Social Sciences: An Agenda for the 21st Century, Amsterdam.

Nadasdy, P. (2003). *Hunters and bureaucrats: Power, knowledge, and aboriginal-state relations in the Southwest Yukon.* Vancouver: UBC Press.

Obomsawin, A. (1984). *Incident at Restigouche.* 45 min 57 s. Canada: National Film Board of Canada.

———. (2002). *Is the Crown at War with Us?* 96 min 31 s. Canada: National Film Board of Canada.

Pictou, S. & A. Bull. (2009). Resource extraction in the Maritimes: Historic links with racism. *New Socialist 1*, 38–39.

Prosper, K., M. J. Paulette, & A. Davis. (2004, August). *Traditional wisdom can build a sustainable future.* Atlantic Fisherman, 2.

Rodríguez, C. (2001). *Fissures in the mediascape: An international study of citizens' media.* Cresskill, NJ: Hampton Press.

Smith, L. (1999). *Decolonizing methodologies: Research and indigenous peoples.* New York: St. Martin's Press.

Veltmeyer, H. (1990). The restructuring of capital and the regional problem. In *Restructuring and resistance from Atlantic Canada*, ed. B. Fairley, C. Leys, & J. Sacouman, pp. 79–104. Toronto: Garamond Press.

Wiber, M. & A. Bull. (2009). Re-scaling governance for better resource management? In *Rules of law and laws of ruling: On the governance of law*, ed. F. von Benda-Beckmann, K. von Benda-Beckmann, & J. Eckert, pp. 151–170. Burlington: Ashgate.

Wiber, M., & J. Kennedy. (2001). Impossible dreams: Reforming fisheries management in the Canadian Maritimes after the Marshall Decision. *Law & Anthropology 11*, 282–297.

Wicken, W. C. (2002). *Mi'kmaq treaties on trial: History, land, and Donald Marshall Junior.* Toronto: University of Toronto Press.

Notes on Contributors

David Austin is the editor of *You Don't Play with Revolution: The Montreal Lectures of CLR James* (AK Press, 2009) and *A View for Freedom: Alfie Roberts Speaks on the Caribbean, Cricket, Montreal, and C.L.R. James* (Alfie Roberts Institute, 2005). His most recent work on the Caribbean left has appeared in *Counterpoints: Edward Said's Legacy*; *New World Coming: The Sixties and the Shaping of Global Consciousness*; *Small Axe: A Journal of Caribbean Criticism*; *The Journal of African-American History*; and *Race and Class*. In addition to his forthcoming *The Unfinished Revolution: Linton Kwesi Johnson, Poetry, and the New Society*, he is currently working on the political thought of the Caribbean Conference Committee, a Montreal-based Caribbean group in the 1960s and its ties to the emergence of the Caribbean New Left. Mr. Austin has also prepared and narrated radio documentaries on the life and work of C.L.R. James (*The Black Jacobins, 2005*) and on Frantz Fanon (*The Wretched of the Earth, 2006*) for the Canadian Broadcasting Corporation's IDEAS.

David Bleakney is a postal worker who since 1996 has been an elected national representative for education for the Canadian Union of Postal Workers; he writes, develops, and delivers courses in collaborative ways for rank-and-file union members. He also served as organization, education, and benefits officer and chief steward at the local level and was a participant at the founding conference of the global network Peoples Global Action. He has designed and conducted workshops in local communities and labor councils around societal and organizational change, popular economics, and direct action, and is active in several international and Indigenous solidarity campaigns. He is a recipient of the Commemorative Medal of Che Guevara from the Central de Trabajadores de Cuba.

Aziz Choudry is an assistant professor in the department of integrated studies in education at McGill University. He has more than two decades of experience working in activist groups, NGOs, and social movements in the Asia-Pacific and North America. A longtime organizer, educator, and researcher with Aotearoa/New Zealand activist group, GATT Watchdog, he also served on the board of conveners of the Asia-Pacific Research Network from 2002 to 2004. Currently he sits on the boards of the Immigrant Workers Centre in Montreal and the U.S.-based Global Justice Ecology Project, and is a co-initiator and member of the editorial team of the collaborative Web site www.bilaterals.org, supporting critical analysis of and resistance against bilateral free trade and investment agreements. He is coauthor of *Fight Back: Workplace Justice for Immigrants* (Fernwood Books, 2009).

Ashwin Desai is based at the Centre for Sociological Research at the University of Johannesburg. He is closely involved with the resurgent social movements in South Africa and has written a book about these developments entitled *"We are the Poors": Community Struggles in Post-Apartheid South Africa* (Pluto Press, 2002). His latest book, coauthored with Goolam Vahed, is *Inside Indenture: A South African Story 1860–1914* (HSRC Press, 2007).

Emma Dowling has been active in summit mobilizations and the organization of the European Social Forum and World Social Forum (official and autonomous spaces), and migrant rights activism for freedom of movement. She researches the transformations and crises of global governance in response to contestation, including how dissent is both disciplined and rendered productive by governance and the questions this raises for social movement activism. Her work also looks at affect in its intersections with gender and labor, as well as politics and knowledge production. She is a lecturer in ethics, governance, and accountability at Queen Mary, University of London.

Wali Haider started his political and social activism as a youth, where his efforts were geared to sensitizing working children in Pakistan toward understanding and agitating against their exploitation. In recent years, much of his work has been to create a platform for political activism with small and landless farmers in Pakistan and to build linkages with regional and international antiglobalization movements. He played an active role in organizing and mobilizing support against the military regime of General Pervaiz Musharraf so that a democratic system of political engagement could be reinstated in Pakistan. Mr. Haider obtained his master's in philosophy from the University of Karachi in Pakistan in 2006.

Adam Hanieh recently completed his PhD in political science at York University in Toronto. He has lived and worked in the West Bank, Palestine, and is active in the Palestine solidarity movement in Toronto. His research is focused on the political economy of the Middle East, and he has published in *Journal of Palestine Studies, Studies in Political Economy, Monthly Review*, and *Historical Materialism*. He is coauthor of *Stolen Youth: The Politics of Israel's Detention of Palestinian Children* (Pluto Press, 2004).

Hsiao-Chuan Hsia is a professor at the Graduate Institute for Social Transformation Studies at Shih Hsin University in Taipei. As the first scholar studying marriage migration issues in Taiwan, her first book is titled *Drifting Shoal: the "Foreign Brides" Phenomenon in Capitalist Globalization* (in Chinese). Her recent academic publications include issues on citizenship and the development of grassroots immigrant and migrant workers' movements. She is also an activist striving for the empowerment of immigrant women and the making of an im/migrant movement in Taiwan. She initiated the Chinese literacy programs for "foreign brides" in 1995, leading to the establishment of TransAsia Sisters Association, Taiwan (TASAT). She is the cofounder of the Alliance for the Human Rights Legislation for Immigrants and Migrants in Taiwan and the Action Network for the Marriage Migrants Rights and Empowerment, an international network for marriage migrant issues. She serves as a board member of the Asia Pacific Mission for Migrants, and is a member of the International Coordinating Body of the International Migrants Alliance.

Kees Hudig is currently working as a freelance campaigner and media activist, mainly on political and economic issues. He has been publishing on activist issues since the early 1980s, both in activist and commercial media. He currently organizes, with others, a monthly debate on "Real World Economics" at the cultural center CREA of the University of Amsterdam, and is affiliated to XminusY Solidarity Fund and Platform DSE (Platform for a Sustainable and Fair Economy). Since 2001 he has been coordinating www.globalinfo.nl, the Dutch Web site on globalization.

Dip Kapoor is an associate professor in theoretical, cultural, and international studies in education in the department of educational policy studies at the University of Alberta, and is a research associate and founding member of the Adivasi/Dalit rural people's Center for Research and Development Solidarity in South Orissa, India. His educational, research, and active engagements with the Adivasi-Dalit Ekta Abhijan (ADEA) land and forest movement began in the early 1990s. Recent edited and coedited collections include: *Education, PAR, and Social Change (Palgrave Macmillan, 2009): International Perspectives*; *Education, Decolonization and Development (Sense Publishers, 2009): Perspectives from Asia, Africa, and the Americas*; *Indigenous Knowledge and Learning in Asia/Pacific and Africa (Palgrave Macmillan, 2010): Perspectives on Development, Education and Culture (Palgrave*

Macmillan, 2009); and *Global Perspectives on Adult Education (Palgrave Macmillan, 2008)*.

Biju Mathew is a cofounder and volunteer organizer with the New York Taxi Workers Alliance, a grassroots mass-based organization of immigrant taxi drivers in New York City. His recent book *Taxi! Cabs and Capitalism in New York City* (Cornell University Press, 2008) bridges the worlds of the academy and labor organizing, and of theory and practice, by combining contemporary social theory with ethnography based in political work. While he continues to work on immigrant labor issues, his current research and political work is also on the emergence and consequences of Special Economic Zones in India. He is associate professor of business at Rider University in New Jersey. He is also a member of the board of the Brecht Forum (The New York Marxist School), cohost of a weekly radio show, "Global Movements/Urban Struggles," on WBAI in New York City, and a cofounder of the South Asian Solidarity Initiative (SASI) and the Forum of Inquilabi Leftists (FOIL).

Michael Morrill is the executive director of Keystone Progress, Pennsylvania's largest and most effective online action network. Keystone Progress helps people engage in direct and online action. He has been involved in organizing since childhood, working for numerous unions, civic engagement organizations, and political campaigns. He has served as executive director of Pennsylvanians United for Quality Care, Pennsylvania's Campaign for Choice, and Pennsylvania Citizen Action. He has written numerous articles and has appeared frequently in national and Pennsylvania media as a spokesperson for working families.

Mario Novelli is a senior lecturer in international development studies at the University of Amsterdam. His broad field of interest explores the relationship between globalization and international development with a particular focus on two themes: labor and education. He has published in journals such as *Globalizations, Globalisation, Societies and Education* and the *Journal of International Educational Development*. He is the coauthor, with Anibel Ferus-Comelo of *Globalisation, Knowledge and Labour* (London: Routledge, 2010).

Sherry Pictou is a coordinator of community programs and the former chief of the Bear River First Nation, a Mi'kmaq community of 150 people on Canada's east coast. She is the North American representative for the World Forum of Fisher Peoples (WFFP), and cochair for the WFFP Coordinating Committee. She is associate staff of the Bay of Fundy Marine Resource Centre and holds an M.A. in adult education from Dalhousie University.

Kumar Prasant is from Mohana village/town area and is convener of the Adivasi-Dalit Ekta Abhijan (ADEA) movement organization in South Orissa, India, and research associate for the Center for Research and Development Solidarity (CRDS), a Dalit-Adivasi people's organization. He has more than twenty years of experience as a local organizer, popular educator, and activist, and has also been the president of the All Orissa Agricultural Labour Union. He is the author and composer of several collected works of political poetry and songs in Oriya that have been instrumental in promoting organizing and activist work in the state of Orissa and in various Dalit national campaigns. He holds bachelor's degrees in education and Hindustani music.

Robyn Magalit Rodriguez is an assistant professor in the department of sociology at Rutgers University. Her first book, *Migrants for Export: How the Philippine Brokers Labor to the World* (University of Minnesota Press, 2010), examines the Philippines' emergence as one of the top labor-exporting countries in the world. Dr. Rodriguez argues that the Philippines is a neoliberal "labor brokerage state" that has made developmentalism a burden and responsibility for the country's worker-citizens. In other publications, she explores how Philippine migrants engage transnationally in counterhegemonic political mobilizations. She explores the ways migrants make claims to rights and belonging undermine the state's project of labor brokerage. Dr. Rodriguez has been actively involved as an immigrant-rights activist and advocate in the U.S.-Filipino community, including the Philippine Forum, a New York–based organization. She has also supported the efforts of the National Alliance for Filipino Concerns (NAFCON), of which the Philippine Forum is part.

Azra Talat Sayeed is a political and social activist working with small and landless farmers and workers in the informal sector. She is executive director for Roots for Equity, a national nongovernmental organization. Dr. Sayeed has also taught at the University of Karachi at the Center of Excellence for Women's Studies. Recently she has been teaching political economy of Pakistan at the Pakistan Study Center, University of Karachi. She earned her PhD at the University of Minnesota in social and administrative pharmacy in 1995.

Martha Stiegman is a Maritime-bred, Montreal-based filmmaker, PhD candidate, and part-time faculty at Concordia University's School of Community and Public Affairs. Her doctoral research and documentary film, *In the Same Boat?* (http://inthesameboat.net, 2007), examines the grounds for solidarity between Mi'kmaq and nonindigenous communities in the fight against the privatization of the Nova Scotia fisheries.

Shannon Walsh holds a PhD from the department of integrated studies in education at McGill University. Her doctoral work examined the rise of neo-liberalism on health strategies in Durban's shack settlements. She has been involved in social movements both in Canada and South Africa over the past decade, and has published in journals such as *AK Press*, the *Review of African Political Economy* (ROAPE), *Feminist Media Studies* and *Sex Education*. She spends the rest of her time as a filmmaker. Her first feature documentary, *H2Oil*, (http://h2oildoc.com, 2009) explores the human and environmental impacts of Canada's tar sands.

Rafeef Ziadah is a PhD candidate in political science at York University in Toronto. Her research examines how Palestinian cultural production is treated within the context of Canadian multiculturalism. She holds an M.A. in international relations from Seton Hall University. Her M.A. thesis analyzed the changes and challenges to the Palestinian women's movement during the years of the Oslo process. She has written on Palestinian refugee rights and on issues related to Palestine exile politics, and released her first CD of spoken word poetry, *Hadeel* (www.rafeefziadah.ca) in 2009.

Index

Abdul (Ahmed Omar Abdikadir), 36–40, 43, 45, 46, 48
academic knowledge, 2, 65, 170
active citizens, 43
activism, 3, 5, 6, 7, 10, 28, 65, 66, 77, 88, 102, 107, 147, 196–197, 205, 208, 220
  see also anti-APEC activism, movement activism, social movement activism
activist(s), 6, 19, 24, 26, 29–30, 66–67, 73–75, 79, 87–89, 93, 95–97, 105, 114, 126, 131–132, 208
  see also feminist activists, migrant activists, political activists, social movement activists
activist ethnography, 23, 158
activist research, 3
Adivasi (original dweller), 5, 11, 193–210
  see also Dalit
Adivasi-Dalit Ekta Abhijan (ADEA), 11, 195–208
African National Congress (ANC), 36, 41, 43, 47
Agamben, Giorgio, 36, 48, 49
agency, 103, 124, 236
  see also collective agency
Albert Park, Durban, 36, 40–43, 49
Aliquippa Medical Center (Pennsylvania), 148

Alliance for Human Rights Legislation for Immigrants and Migrants (AHRLIM), 102, 105, 106, 111
al-Nakba (catastrophe), 85–86, 89, 91, 92
alternatives to globalization, 18
Alvarez, Sonia, 59
Amnesty International, 132
anarchists, 72, 74
Anjuman-e Mazareen Punjab (AMP, Pakistan), 11, 211–225
antagonism, 69–83
anti-Apartheid movement, 85–99
anti-APEC activism, 7, 23
  see also APEC
antiglobalization movement, 19, 244
  see also counter-globalization movement(s), global justice movement
anti-privatization struggles, 121
Aotearoa/ New Zealand, 5, 26, 29
Apartheid analysis, 87, 89, 98
"Apartheid Wall" (Israel), 85, 90, 98
Arellano, Elvira, 62, 67
Armstrong, Lisa and Prashad, Vijay, 22
Asia-Pacific, 5
Asia-Pacific Economic Cooperation (APEC), 23–25, 27–28, 32
  see also anti-APEC activism
Austin, David, 10, 173–189, 243
Ayala Group of Companies (AGC), 57

Barker, Michael, 77
Bear River First Nation (BRFN) (*L'setkuk*), 11, 227–242
Bevington, Doug and Dixon, Chris, 4, 6
Biel, Robert, 23
bilateral free trade agreements, 5
Bishop, Maurice, 176, 179, 181–183, 187
Black power movement, 176
Bleakney, David, 9, 139–155, 243
Boal, Augusto, 106, 108, 112
Boycott, Divestment, and Sanctions (BDS, Palestine), 88–98
British colonial rule, 212
British Trades Union Congress (TUC), 121, 133
Burgmann, Verity, and Ure, Andrew, 26
Burnt Church (Esgenoôpetitj), 231–232
Burrow, Sharan, 59, 63
Bustamante, Jorge, 64–65

Cali, Colombia, 9, 121, 127, 134
Canada, 5, 8, 9, 11, 26, 28–30, 90, 97, 133, 139–142, 145, 149–151, 177, 227–242
Canadian Union of Postal Workers (CUPW), 140, 149–150
capitalism, 7, 19–21, 23, 25, 47, 75, 89, 122, 158, 199, 228
see also global capitalism, market capitalism
capitalist globalization, 17, 25–26, 29–30, 101, 104, 111
Caribbean, the, 173–189
Caribbean left, 10, 175, 176, 182, 185
caste, 5, 193–210
-communalism, 11
system, 194
casteism, 197
Celeyta, Berenice, 132, 134
Center for Research and Development Solidarity (CRDS), 132, 134
Choudry, Aziz, 1–13, 17–34, 243
civil disobedience, 73, 221, 230

"civil society"
meetings, 8, 31, 53–54, 58–66
process(es), 8, 53, 64–65
Coard, Bernard, 176, 179, 181, 183, 187
coastal communities, 228, 231–232, 234–235
Cold War, 174–175, 185
collective agency, 9, 103
collective ethnographies, 157, 168, 170
collective(ist) approaches to activist knowledge, 8, 85–99
Collins, Merle, 173, 176, 186
colonialism, 3, 7, 18, 20, 25, 26, 29–30, 86, 175, 185, 228
Colombia, 9, 121–137
communal subject, 107–109, 111
see also historical subject, personal subject
compartmentalization, 18, 22
Council of Canadians, 27–29
counter-globalization movement(s), 69, 70, 74, 78, 81
counterhegemonic narrative, 87
critical consciousness, 3, 104, 123
critical learning, 3

Dalit, 5, 11, 193–210
see also Adivasi (original dwellers)
Davos, Switzerland, 71
Department of Fisheries and Oceans Canada (DFO), 227–242
Desai, Ashwin, 6, 8, 35–51, 244
Desai, Bhairavi, 157
deserving citizen discourse, 46
Desmarais, Annette, 24, 31
development, 2, 11, 19–21, 54, 64
dialectics, 2, 124, 177
diversity, 29, 71, 124, 143–144, 234
documentary video, 107, 227–242
documentation, 6, 206, 212
domination, 1, 11, 24, 46, 123, 224
Dowling, Emma, 8, 69–83, 244
Dutch Social Forum, 76–77

EMCALI, 121, 126–130, 132
  see also SINTRAEMCALI
empowerment, 9, 23, 102, 104,
  106–107, 109, 112, 115
epistemologies of knowledge, 2, 3, 31
Europe, 2, 8, 39, 70, 72, 74, 102, 104
European Social Forum (ESF), 8,
  74–75, 80, 133
expertise, 8, 53–68

Fanon, Frantz, 186
farmers, 24–25, 29–31, 101, 125, 150,
  168, 177, 204, 211, 213,
  216, 220
feminist activists, 9
fisheries, 11, 227–242
flexible subsumption, 164, 167–169
Foley, Griff, 3–4, 23, 112, 114, 124
forced migration, 11, 197, 202–203
"foreign brides", 9, 103, 105–106
formal subsumption, 165–167
forum theater, 108, 110, 112
Freire, Paulo, 104, 106, 108, 123,
  128, 144

G8, 71, 76–78, 149
Gairy, Eric, 174–175, 178–181, 185
gatekeeping, 21, 31
Gaza Strip, 86–89, 92, 99
genealogies of intellectualism, 158
General Agreement on Tariffs and
  Trade (GATT), 24, 244
global capitalism, 7, 23, 24, 26, 30, 70
  see also capitalism
Global Forum on Migration and
  Development (GFMD), 8,
  53–68
Global Forum on Modern-day Slavery,
  53, 62
globalization, see capitalist
  globalization, neoliberal
  globalization
global justice, 17, 22, 23, 26, 31
global justice movement, 7, 67
global North, 2, 70, 71

global positioning system (GPS), 10,
  159–161, 164–168
  see also Taxi Technology
  Enhancement Program (TTEP)
global South, 2, 71, 139
grassroots migrant activists, 8, 53
Grenada, 10, 173–189
Grenada Revolution, 10, 173–189
Gupta, Rahila, 6

Hage, Ghassan, 29
Haider, Wali, 11, 211–225, 244
Hanieh, Adam, 6, 8, 85–99, 244
Hart, Richard, 182–183, 185
Harvey, Franklyn, 177–179, 185, 187
Hector, Tim, 180, 184, 185
hegemonic positions, 30, 31
hierarchies of knowledge production, 5,
  23, 30–32
Hill, Robert, 177
Hindutva, 11, 203–205
historical subject, 103, 107, 109, 111,
  114, 116
  see also communal subject, personal
  subject
historicity (Touraine), 103, 114
history, 6–7, 28, 43, 47, 61, 86, 91, 103,
  104, 139, 151, 166, 185, 212, 213,
  215, 218
  critical approaches to, 6
Holst, John, 4
horizontal (self-organized) networks,
  8, 70, 74
Hsia, Hsiao-Chuan, 9, 101–118, 245
human rights, 25, 36, 47–48, 75, 88,
  104, 132–135, 195, 207
Human Rights Watch (HRW), 132, 219
Human Sciences Research Council
  (HSRC), 44–46
Hudig, Kees, 8, 69–83, 245

ideology of pragmatism, 7, 18, 20, 21
im/migrant movement, 9, 102,
  104–106, 108, 113–114
imperialism, 3, 7, 20, 22, 75, 175, 228

incidental learning, 3, 4, 124
India, 5, 11, 168, 193–210
Indigenous Peoples, 5, 24, 26–31, 91, 139, 145, 146, 227, 228, 235, 241
Indigenous struggles, 10, 91, 96
individual dislocated person, 208
Individual Transferable Quotas (ITQs), 231–232
industrial encroachments, 202
informal learning, 2, 4
In the Same Boat? (film), 228, 235–240, 247
Inter-American Court of Human Rights (IACHR), 132, 134
International Assembly for Migrants and Refugees (IAMR), 53, 62–66
International Migrants Alliance (IMA), 63, 66, 115
International Monetary Fund (IMF), 70, 71, 79, 122, 127
International Trade Union Confederation (ITUC), 58, 59, 63
Israel, 85–90, 97–99
Israeli Apartheid Week, 86, 93

Jackson, Moana, 28
Jad, Islah, 59
James, C.L.R., 177–179, 185

Kamat, Sangeeta, 18, 19, 20
Kananaskis, Canada (G8), 149
Kandhamal, India, 196, 197, 202–207
Kapoor, Dip, 1–13, 193–210, 245
Kelley, Robin, 4
Khanewal, Pakistan, 211–213, 221–222
Khoza, Vusi, 41–43
Kinsman, Gary, 3, 23
knowledge
    industry, 35
    politics, 5, 17
    production, 1–13, 23, 30, 36, 48–49, 64, 66, 67, 70, 75, 91, 94–95, 102, 123–124, 129, 139, 143, 150, 157–159, 168–170, 193, 207–208

    see also sustainable knowledge sharing, traditional knowledge, traditional ecological knowledge

labor education, 9, 10, 142, 144
labor subsumption, 166
Lamming, George, 184
land and forest alienation, 11, 195–197, 201, 205, 207
land/forest classification schemes, 11, 196, 198, 201
land rights movement, 212, 214, 224
Latin America, 7, 59, 123, 133, 180
Lawyers for Human Rights (LHR), 41–42, 49
learning, *see* critical learning, formal learning, incidental learning, informal learning, learning in struggle, learning through action, nonformal learning, strategic learning, trade union learning
learning in struggle, 5
learning through action, 227
left party-political activism, 9, 119
Lestari, Eni, 63, 67
liberation, 10, 36, 92, 96, 97, 174, 182
lobbying, 20, 21, 146
local communities, 46, 121–122, 127–128, 228
Lok Adhikar Manch (LAM), 207–208
Lopez, Alexander, 126–129

Madondo, Eugene, 36, 40–43, 46, 48, 49
macro-micro analyses, 3
"Make Poverty History" (MPH), 77–78
Manila, Philippines, 8, 53–59
Maori, 2, 26, 28, 228
market capitalism, 29
marketization discourse, 24
marriage migrants, 9, 101–118
Marshall (Decision), 227–228, 231–235
Marshall, Donald, Jr., 230
material theory, 168

Marx, Karl, 103, 165–166, 168–169, 184
Marxist-Leninist theory, 10, 174
mass-based organizing, 6
Mathew, Biju, 7, 10, 157–171, 246
McNally, David, 19, 20
media solidarity, 217
Meeks, Brian, 175, 179, 181
Mertes, Tom, 1
Mignolo, Walter, 2
Migrante-International, 56, 61, 67
migration, 11, 45, 54, 55, 57–64, 197, 199, 202
migration as development, 54, 62
migrants, 8, 35–37, 41, 43–49, 53–67, 139, 146
see also marriage migrants
Mi'kmaq, 11, 227–242
militant activism, 221
military farm management, 211–218, 220–221
see also Pakistan military farm
mobilization(s), 2, 6, 7, 8, 10, 19, 21, 26, 30, 41, 47, 56, 63, 70, 71, 73, 88, 130, 163, 168, 215, 217, 218, 219, 232
Morrill, Michael, 9, 139–155, 246
movement activism, 1, 11, 18, 205
movement
-building, 93, 95, 145
-centric knowledge and learning, 2
-relevant theory, 4
movement(s), *see* anti-Apartheid movement, Black power movement, counter-globalization movements, im/migrant movement, land rights movement, Palestine solidarity movement, social movements, solidarity movement
movement of movements, 69, 71
multiculturalism, 29, 107

National Rural Employment Guarantee Scheme (NREGS, India), 200–201, 206

neoliberal discourse, 24
neoliberal globalization, 23, 25, 27, 29, 62, 65, 69, 71, 135, 228, 229
neoliberalism, 10, 18, 19, 25, 26, 29, 69–70, 77, 123, 125, 153, 164
Netuklimk, 227–242
New Jewel Movement (NJM, Grenada), 174, 178–180, 183
Newtown, Johannesburg, 37, 40, 49
New York City (NYC), 157–159, 170–171
New York Taxi Workers Alliance (NYTWA), 10, 157, 159–160, 162–163, 165, 168–169
nongovernmental organizations (NGOs), 1, 5, 8, 17–34, 35, 44, 46, 48, 54, 57–60, 62–65, 74–81, 102, 125, 222–223
as intellectual policemen, 21
careerism, 6
ECOSOC status, 58–59
NGOism, 19, 26
NGOization, 7, 17, 19, 22, 205
nonindigenous fishers, 11, 227–228, 231–234
nonindigenous fishing community(ies), 227, 231, 237
nonprofit industrial complex, 6
nonsectarianism, 9, 93
nonsectarian movement building, 87, 91, 93
Nova Scotia, Canada, 11, 227–234
Novelli, Mario, 9, 121–137, 246

organizing work, 10, 30, 157
Okara, Pakistan, 211–225
oppressed peoples, 4
organizing knowledge, 21
Orissa, India, 5, 11, 193–210
Oslo process (Palestine), 87, 89, 98
"Ownership or Death", 11, 211, 214, 215, 222
Oxfam, 25, 78–79

Pacific, the, 2
Pakistan, 11, 211–225

Pakistan military farms, 11
  see also military farm management
Pakistan Seed Corporation (PSC), 11, 211–214
Palestine, 8, 59, 85–99
  solidarity movement, 8, 86, 88, 90, 93
Palestinian narrative, 87, 91–94, 97
parliamentary politics, 179, 185
participation, 4, 19, 20, 96, 102, 105, 115, 220, 223, 231, 241
  see also popular participation
participatory video, 236
Peace and Friendship Treaties (Canada), 227, 229, 230, 233
peasants, 202, 212–215
pedagogy of mobilization, 4
Peoples' Global Action (PGA), 61
Peoples' Global Action on Migration, Development, and Human Rights (PGAMDHR), 61, 64, 66
People's Revolutionary Government (PRG), 175, 179–183
People's Summits, 26
personal subject, 103, 107, 108, 109
  see also communal subject, historical subject
Petras, James and Veltmeyer, Henry, 20, 21
Philippines, 24, 56, 57, 60, 101
Phillip, Nicole, 176, 180–182
Pictou, Sherry, 11, 227–242, 246
Plan PARE (Colombia), 128–131, 134
policy analysis, 24, 31
political activists, 122, 123, 133, 134, 222
political activist ethnography, 23
political education, 2, 30, 94, 96, 131
political mobilization(s), 10, 19, 65
political work, 10, 78, 123, 158–159, 169, 241
politics of
  division, 11, 196–198, 202, 205, 207
  domination and resistance, 1

knowledge production, 7, 8
popular education, 5, 24, 123–124, 146, 205
popular participation, 180, 182, 183
  see also participation
post-Apartheid, 35, 47
poverty, 11, 54, 64, 78, 101, 195, 201, 205, 212
  see also "Make Poverty History"
power dynamics, 3, 73, 148, 180
power struggles, 74
pragmatism, *see* ideology of pragmatism
Prague, 71, 73, 79
Prasant, Kumar, 10, 193–210, 247
privatization, 9, 18, 70, 121, 122, 125, 126, 129, 130, 231
  of fisheries, 11, 234–236
  of resources, 227–228, 231
professionalization, 8, 19, 69, 77, 79, 81
  of activism, 77
  of social change, 17, 32
professionalized intellectual policemen, 21
protest camps, 73–74
public intellectual, 169–170
public services, 9, 127–131, 134
Punjab Seed Corporation (PSC), 211–214

real subsumption, 166–167
refugees, 8, 36, 38, 39, 43–46, 48, 87, 89–92, 98
resistance, 1, 8, 9, 11, 23, 25–27, 30, 47, 65–66, 103, 122, 123, 125, 135, 144, 146, 147, 175, 204, 216, 221, 228, 230, 231
  see also zones of resistance
resource exploitation, 228
Rodriguez, Dylan, 17, 20
Rodriguez, Robyn Magalit, 6, 8, 53–68, 247
Roopnaraine, Rupert, 185–186

Saigol, Rubina, 212, 215
Sangh Parivar, 203–204

Santo Tomas, Patricia, 60
Sayeed, Azra Talat, 11, 211–225, 247
Scheduled Caste (SC), 194–195, 197–198, 201
Scheduled Tribes (ST), 194–199, 201
scholar-activists, 116
Second Intifada, 87, 88, 97
sectarianism, 93, 94
self-determination, 26–29, 87
Service Employees International Union (SEIU), 140, 147–148
sharecroppers, 11, 212
SINTRAEMCALI, 9, 121–137
see also EMCALI
Sinwell, Luke, 47
slavery, 193, 213
Smith, Andrea, 21
Smith, George W., 3, 23
Smith, Linda T., 3, 22, 228
social change, 1, 2, 5, 10, 17, 20, 32, 59, 71, 75, 123, 140, 177
social forums, 72, 74, 75, 81
see also Dutch Social Forum, European Social Forum (ESF), World Social Forum (WSF)
socialism, 10, 174, 175, 179
Socialist Workers Party (SWP), 73–74
social justice struggles, 3
social movement(s), 1–13, 17, 21, 23, 25, 30–32, 43, 46–48, 59, 65, 66, 75, 79, 103, 107, 114–116, 122–126, 132, 139–140, 144–147, 150–151, 174–175, 193
social movement
  activism, 8, 70, 75, 81, 207, 244
  activists, 21
  education, 150, 152
  learning, 139
  scholarship, 1
  theory(ies), 4, 103, 114
social struggles, 2, 3, 5, 151
social transformation, 6, 10, 18, 30, 174, 177, 178, 180, 182, 184

societal movement, 103, 115
solidarity, 5, 8, 9, 25, 29, 115, 121, 122, 133, 150, 195, 197, 207, 222, 228, 239, 240
  networks, 132, 134
  see also media solidarity, Palestine solidarity movement
Somalia, 36, 37, 39
Somalia Association of South Africa (SASA), 39
South Africa, 8, 35–51, 88–90
Southeast Asia, 9, 101–102, 108
Special Economic Zones (SEZs), 11, 196, 197, 202–204, 206
Stiegman, Martha, 11, 227–242, 247
strategic learning, 122, 124, 125, 135
strategy development, 121–122, 124, 135
strike(s), 10, 121, 131, 157, 160, 174, 206, 221
subjectivation, 9, 101–103, 107, 109, 111, 115–116
subsumption, *see* flexible subsumption, formal subsumption, labor subsumption, real subsumption
summit protests, 8, 70, 72, 81
sustainable knowledge sharing, 87, 91, 95

Taiwan, 9, 101–118
taxi(cab) drivers, 10, 157
taxi industry, 159, 164–167, 170
Taxi Technology Enhancement Program (TTEP), 159
  see also global positioning system (GPS)
Taylor, Jeff, 141–142, 145
temporary labor migration programs (TLMPs), 53–56, 59–60, 63–64
third world, 3, 7, 29, 30, 31, 157
Toronto, Canada, 8, 30, 85–99, 140, 176
Touraine, Alain, 9, 102–103, 107, 109, 114, 115, 116

trade unions, 9–10, 27, 55, 58, 72, 76, 77, 122, 123, 139, 142–143, 177
see also British Trades Union Congress, Canadian Union of Postal Workers (CUPW), Canadian Union of Public Employees (CUPE), International Trade Union Confederation, Service Employees International Union (SEIU), SINTRAEMCALI, union education, unions
traditional knowledge, 11, 24, 234, 238
traditional ecological knowledge (TEK), 239
TransAsia Sisters Association (TASAT), 105–108, 110–115
translocal activism, 6, 207
Trotskyists, 73

union bureaucracies, 141
union/community alliance, 127, 128, 134
see also worker/community alliance
union education, 10, 122, 140–143, 145, 147, 150–151
see also trade unions
unions, 86, 90, 95, 123, 140–142, 144–148, 151
see also trade unions
United States, 9, 26, 39, 87, 104, 133, 139–142, 145, 147–148, 151, 161, 170, 174–175

vanguardism, 8, 69, 74, 178, 179, 185
vanguardist politics, 10
Vedanta Alumina Ltd., 202
video, see documentary video, participatory video

violence, 8, 11, 35–40, 43–49, 72, 90, 125, 196, 202, 206, 208, 232
voice, 62, 91, 92, 147, 236, 238
"volleyball" practices in organizing, 113–114

Walsh, Shannon, 6, 8, 35–51, 248
Watkins, Kevin, 25
West Bank, 85–89, 92, 98, 99
"white progressive economic nationalist", 27
Wolpe, Harold, 46
Wood, Ellen Meiksins, 18–19, 21
worker alliances, 9
worker/community alliance, 127
see also union/community alliance
worker education, 139–140, 142, 144–148
worker education practices, 151
working-class, 65, 101, 139, 143, 150, 178
World Bank, 54, 71, 79, 122, 127
World Social Forum (WSF), 71–72, 75–77, 81, 82, 133
World Trade Organization (WTO), 7, 23, 24, 70, 79

xenophobia, 35–37, 39, 43, 44, 46, 48

Ziadah, Rafeef, 6, 8, 85–99, 248
Ziai, Aram and Schwenken, Helen, 54, 55
Zimbabwe, 40–43, 49
Zionism, 89–90, 92, 97
zones of resistance, 9
see also resistance